The U.S.-Honduras Remittance Corridor

Acting on Opportunities to Increase Financial Inclusion and Foster Development of a Transnational Economy

Isaku Endo
Sarah Hirsch
Jan Rogge
Kamil Borowik

THE WORLD BANK
Washington, D.C.

On behalf of
Federal Ministry
for Economic Cooperation
and Development

ISBN-13: 978-0-8213-8139-7
eISBN: 978-0-8213-8146-5
ISSN: 1726-5878   DOI: 10.1596/978-0-8213-8139-7

Library of Congress Cataloging-in-Publication Data has been requested.

Contents

Figures

Boxes

Foreword

As the current financial meltdown continues to rage globally, remittance flows by migrant/guest workers from the United States to their home countries have not gone unaffected; and remittance flows to Honduras are no exception.

This report analyzes the U.S.-Honduras remittance corridor and builds on lessons learned from international experiences on remittances. It was prepared with extensive interviews of government authorities, financial regulators, market participants, Honduran migrant communities, NGOs, and local communities receiving remittances.

It also highlights critical policy recommendations for authorities to improve the integrity of the remittance flows; expand access to financial services; and create an environment where Honduran migrants in the United States can invest in their community and link diaspora groups and home communities.

Six areas provide the focus of this report: (i) regulatory reforms for the remittance market are urgent in order to improve clarity in regulations as well as to include money transfer companies in the regulatory framework; (ii) money service businesses would benefit from an examination of state regulation and their subsequent harmonization and coordination; (iii) rural areas need to improve distributive infrastructure to better reap the benefits of the remittance flows; (iv) financial education and awareness for Honduran migrant communities are critical components with the overall remittance flow equation; (v) the regulatory environment of remittance flows would be greatly enhanced through the promotion, inclusion, and expansion of proper identification; and (vi) public policies can be directed to building an environment for diaspora's investments in the community and local business developments for exports to Honduran communities abroad.

This report is a result of the collaborative efforts between the Financial Market Integrity Unit of the World Bank (FPDFI) and the Deutsche Gesellschaft für Technische Zusammenarbeit (GTZ) in Honduras on behalf of the German Federal Ministry for Economic Cooperation and Development to bring together different expertise on remittances, migration, and economic development. FPDFI developed a methodology for research on Bilateral Remittance Corridor Analysis (BRCA) and has applied it to a series of BRCA studies. GTZ in Honduras has conducted research on migration, transnational bridges, and the impacts of remittances.

Hopefully, the dissemination of this report will promote public discussion and lead to solutions that will benefit the Honduran people.

Consolate K. Rusagara
Director
Financial Systems Department
Financial and Private Sector Development
The World Bank

Wolfgang Lutz
Country Director
Deutsche Gesellschaft für Technische
Zusammenarbeit (GTZ)—Honduras

Acknowledgments

This report is part of a series of Bilateral Remittance Corridor Analysis projects undertaken by the Financial Market Integrity Unit of the World Bank. The report was jointly conducted by GTZ in Honduras and the World Bank. The authors of the report are Isaku Endo, Sarah Hirsch, Jan Rogge, and Kamil Borowik.

We are grateful for the comments, guidance, and encouragements from Latifah Merican Cheong, Jean Pesme, Christian Königsperger, Raul Hernández-Coss, and Jürgen Popp. Special thanks go to Adrian Fozzard, Dante Mossi, Massimo Cirasino, Mario Guadamillas, and Peter Feldmann for guidance and consultations. Peer reviewers for this report were Andrea Riester, Imke Gilmer, and Hans Schimpf (GTZ) and Humberto López, Dante Mossi, and José Antonio Garcia Garcia (World Bank). The final text benefited from Sheldon Lippman's editorial advice.

The authors are thankful for support in arranging missions from country offices in Tegucigalpa, including from Martha Magermans, Karla Cerrato, and Nadia Raudales (GTZ) and Eva Melisa Caballero, Ana Funes, Iris Medina Hernandez, Carol Mejia, and Noris Salinas Reyes (World Bank).

Finally, we thank the following people for their helpful suggestions, comments, and valuable information: Luis Agurcia, Fernando Agurcia, Armando Busmail, Celeo Alvarez Casildo, Maricruz Aparicio de Sánchez, José Luis Arita, German Asdrubal, José Francisco Avila, Mario Avila Gutierrez, Jorge Bueso Arias, Armando Castañeda, Marco Caceres, Jimena Calderón, Patricia Canales, Michael Casparian, Patricia Castillo, Nancy V. Castillo Figueroa, Celso Castro, Raul Cerna, Rosario Cobar, Mirtha Colón, Alan Cox, Hugo Cuevas-Mohr, John Dinin, Juan de Dios, Christopher Duque, Fernando Escoto, Violeta Flores, Betsabé Franco, Shin Fujiyama, Jimena García, Ramón Augustin García, Isaac Gorena Espinoza, Roy Guevara, Colón Angel Hamilcar, René Herrera, Raquel Isaula, Fernando Izaguirre, Hiroshi Kawano, Jose Lagos, Xiomara Lurdes Lara, Lina Martínez, Renán Marquez, Jose Marquina Santos, Fabio Matute, Gabriel Matuty, Alex Mayr, Javier Medina, Wilfredo Medina, Jossi Mejía, Ely Melendez, Erica Narvaez, Miguel Navarro, Alejandra Osario, Rodolfo Pastor de Maria Campos, Gloria Jesús Pérez, Francisco Portillo, Patricia Rodriquez, Reynieri Rodriguez, Tania Sagastume de Bueso, Bayardo Salgado, Jose Marquina Santos, Kai Schmitz, Gabriel Sierra, Angelo Sigismondo, Regina Stone, Tony Stone, Pedro Torres, José Leonel Valladares, Peter Vandivier, Manuel Antonio Villa, and Edith Zavala.

Executive Summary

This report on the U.S.-Honduras remittance corridor describes the remittance regulatory and market environment, financial inclusion strategies by financial institutions, transnational economic activities, and the impacts of remittances on the Honduran economy.

In 2008, the environment surrounding remittances dramatically changed along with the deteriorating economic situation spreading across the globe. The year began with an existing weak U.S. dollar, high oil prices, and a housing sector crisis caused by risky subprime mortgages. The U.S. financial downturn immediately spread into an international financial crisis, resulting in slowing economic growth on a global scale. Remittances were no exception to the negative impact of the financial crisis as an economic slowdown in migrant host countries affects employment and incomes.[1] Consequently, the current financial crisis impacts negatively on remittances for Latin American countries, including Honduras, whose incoming remittances are mainly from the United States.

Rapid changes in the remittance environment have had implications in the preparation of this report, a joint effort of GTZ and FPDFI of the World Bank. Although the authors tried to include updated information in the report, the fast-changing economic conditions in the world have made this difficult to achieve. Bringing together local and international knowledge of remittances and applying BRCA methodology, the report focuses on relevant public policy issues for remittances and related matters such as access to finance, regulation, the essence of the remittance market, and community initiatives (transnational bridges). The study missions in the United States and Honduras were undertaken in April 2008.

Overview of Migration and Remittance Trends

According to the World Bank, recorded remittances to developing countries are estimated to reach US$305 billion in 2008, despite a sharp slowdown in growth in the third quarter.[2] Remittances to the Latin American and Caribbean (LAC) Region appear to have experienced zero growth rates in 2008.

Honduras is a relatively large remittance-receiving country in the LAC Region. In 2008, in absolute volume, Honduras received US$2.8 billion in remittances. In the previous year, remittances to Honduras accounted for 21.3 percent of its gross domestic product (GDP). Reflecting migration statistics, 91.4 percent of remittance senders were in the United States. At the household level, remittances constitute the third largest source of household income in Honduras and are largely used to finance basic living expenses.

Figure 1. Remittances and Capital Flows to Developing Countries

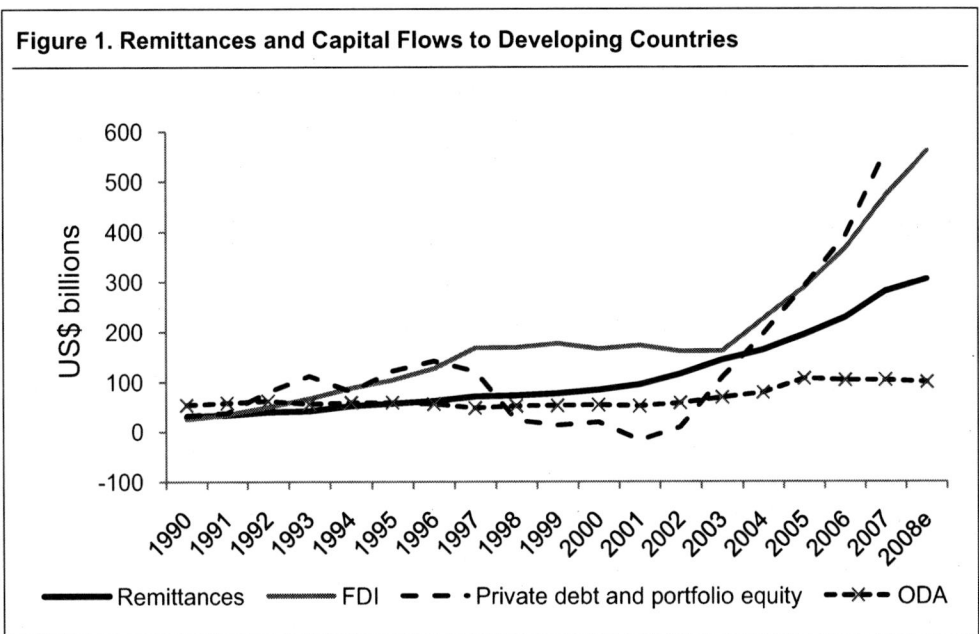

Source: The World Bank—Development Economics Prospects Group

The majority of Hondurans migrate abroad for economic reasons. According to a study by Instituto Nacional de Estatística (INE), 91 percent of Honduran migrants emigrated to seek jobs. At the same time, migration is also triggered by the income differentials and the wage gap of 9 to 12 times for unskilled labor. This shows that potential migrants find existing jobs unsatisfying with regard to income and seek better income opportunities through migration. The principle pull factor for recent migration to the United States has been the booming construction industry, absorbing 47.9 percent of Honduran male migrant labor.

Strong social networks between migrants and their relatives support and facilitate Honduran migration. Many migrants borrow money from family members or friends when they migrate to the United States. There are wide ranging estimates of Honduran migrants in the United States. The American Community Survey (ACS) 2007 estimates a foreign (Honduran-born) population of 430,504 in the United States while the INE estimates 232,069 Honduran emigrants in 2006. The Central Bank of Honduras (BCH) estimates no less than 10 percent of the total population of Honduras or 730,000. The largest five U.S. destination states for Hondurans are Florida, New York, California, Texas, and New Jersey. But the recent U.S. economic slowdown has forced new and established migrants to pursue opportunities in other states.

Notably, return migration represents a significant source for local development in Honduras, yet its active promotion is overshadowed by the increasing deportations. Although the development impact of return and cyclical migration on society and economy in Honduras is not fully evident, three main patterns of return migration have been observed: voluntary temporary and cyclical migration, deportation, and temporary labor programs. Honduran migrants do not necessarily intend to stay permanently in the United States; some plan to return home or make frequent return trips to the United States, thus initiating a cyclical migration scheme.

Although remittances to Honduras increased to unprecedented amounts in absolute number, a marked slowdown in the growth rate occurred in 2005–07. The downward trend in the growth rate of remittances is most likely explained by the slowdown in the U.S. economy and tightening of U.S. and Mexican border controls. In addition, it is possible that remittances were over-reported when a new method of data collection was put in place and adjustments were subsequently made. According to the Honduran authorities they have begun dialogues with neighboring states to exchange experiences on migration and remittances; they also stated that they adopted a national policy for emigrants.

The U.S.-Honduras Market for Remittances

Migrants' choices of remittance channels are influenced by socioeconomic, cultural, and institutional reasons and by their migration status. Convenience, cost, and location seem to be the major factors in determining choice of remittance channel. Money transfer companies, a preferred channel, offer remittance services that meet Honduran migrants' needs. As a result, about 92 percent of remittances in the U.S.-Honduras corridor are transferred through formal (regulated) remittance service providers. Honduran migrants in the United States use primarily large money transfer operator (MTO) networks.

In Honduras, the remittance market is highly concentrated among banks with recent expansion to microfinance institutions. Still small, the microfinance institution market is finding its niche. International money transfer companies are also in the remittance market. Despite a growing network, availability of remittances services in rural areas is limited. Credit and savings cooperatives bring access to remittance services in rural Honduras. In 2006, cooperatives distributed about 20 percent of all remittances sent to rural areas. Struck by security issues, further expansion of the network of remittance-paying agents is limited.

Costs of sending remittances to Honduras are low, but not the lowest when compared to those of other corridors between the United States and countries in Latin America. When sending and claiming a remittance in the U.S.-Honduras remittance corridor, the primary associated cost is from the commission paid by sender at origination. These costs are distributed among the capturing agent, intermediaries/network, and distributing agent. Different pricing schemes by remittance service providers in the corridor depend on partnerships and destinations.

The Impact of Regulations on Remittance Markets

Commercial banks and money service businesses operate as remittance service providers in the U.S. remittance market. The U.S. remittance market has regulations at both state and federal levels. State regulators cover the operations of state-chartered banks and money service businesses. Each state has different requirements for licensing in spite of ongoing efforts by regulators to harmonize state regulations.

The Federal Government regulates federal-charged banks and money service businesses on issues of anti-money laundering and combating the financing of terrorism (AML/CFT). The Financial Crimes Enforcement Network (FinCEN) is the U.S. financial intelligence unit, a bureau of the U.S. Department of the Treasury, and administrator of the Bank Secrecy Act. At the federal level, remittance service providers are required to file reports on suspicious activity and currency transactions above certain thresholds. Banks are required to file a suspicious activity report (SAR) on transactions or attempted transactions of at least US$5,000 if the bank knows, suspects, or has reason to suspect money-laundering activities. Money service businesses are required to report transactions or attempted transactions involving at least US$2,000. All remittance service providers are required to file a currency transaction report on transactions in excess of US$10,000. In all cases, remittance service providers must conduct customer identification and verification and can accept government-issued identification, including at their discretion foreign government-issued identification.

In Honduras, the National Commission of Banks and Insurance (Comisión Nacional de Bancos y Seguros or CNBS) has legal authority to supervise financial institutions. The Central Bank of Honduras (Banco Central de Honduras or BCH) is responsible for the oversight of national payment systems and for foreign exchange regulations. The 2002 AML Law established la Unidad de Información Financiera (UIF), Honduras' financial intelligence unit. The AML Law requires supervised and other relevant financial institutions to establish formal AML policies and procedures, including appointing a Compliance Officer and Compliance Committee, know-your-customer (KYC) policies and procedures, ongoing monitoring of customers, and filing suspicious transaction reports (STRs) to the UIF. Under the Honduran legal and regulatory framework, money transfer companies are regulated under the AML/CFT regime but not regulated as a financial institution by either CNBS or BCH.

With respect to KYC requirements for financial institutions, it seems to be unclear whether a client's physical presence is needed at the time of opening an account or making a transaction. The AML Law prescribes customer identification; however, it does not require physical presence. At the same time, authorities interpret that physical presence is necessary although this interpretation is not publicly issued. A few financial institutions in Honduras will open accounts for Honduran migrants while they are in the United States.

Remittances and Financial Inclusion

Massive flows of remittances present a historic opportunity for Honduras to upgrade its financial sector and increase financial inclusion of the poor. Financial inclusion refers to giving people who formerly had no access to formal financial systems access to financial services such as accounts, credits, and insurance products. Empirical studies across the globe suggest that development of the financial sector and financial inclusion has a positive impact on economic growth. There is potential for further financial inclusion of remittance recipients in Honduras.

Remittances to Honduras increase bancarization of remittance recipients although at low overall levels. According to a survey commissioned by the Inter-American Development Bank (IADB), remittance recipients are interested in accessing more financial services. But, Honduran migrants in the United States face obstacles in accessing a range of financial services, starting with lack of documentation. Most migrants do not hold a valid U.S. entry visa or a U.S. social security number, and often they either have lost their Honduran identification in transit or were too young to hold any form of Honduran identification when they left the country. Financial institutions in Honduras have recognized the opportunity for financial inclusion of migrants in the United States and remittance beneficiaries in Honduras. The predominant use of formal remittance channels creates an amicable environment for financial inclusion. Financial institutions in Honduras have adopted marketing strategies to turn remittance senders and receivers into banking clients. These methods depend on the general attitude of a financial institution toward the market, the level of available information, the use of technology, and regulatory aspects in Honduras and in the United States. Most financial institutions in Honduras focus on the receiver as the gateway to financial inclusion. Banks and other financial institutions are positioned to increase financial education as a precondition to greater financial inclusion. Several financial institutions have pioneered the use of online banking and mobile phones to expand access to services by receivers.

Credit and savings cooperatives offer particular benefits for development at local level. These institutions develop lending products tailored to specific development needs of local communities. Some cooperatives have introduced a special line of products for remittance receivers called UNIREMESAS and offer many individual services to migrants based on their knowledge of local communities.

Development Impact of Remittances in Rural Honduras: Transnational Economy, Networks, and Diaspora Engagement

Honduran migrants are partners in the social, economic, and political development of their home communities. A rising transnational economy in rural Honduras can be characterized by migrants' financial contribution to community development, returning migrants and their investments in local private sector, courier services, and informal market of migration and remittances. Migrants in the United States create demand for a market in nostalgic products from Honduras. Viajeros (couriers) deliver specialized, transnational, door-to-door, export-import courier services between certain regions in Honduras and their correspondent migration networks in the United States.

Self-organized migrants promote community development through highly scattered collective remittance initiatives, contributing to their home society through investment and skill transfers. As a good practice example, migrants from Intibucá and Olancho return and invest their savings and new skills in local business. In the town of La Esperanza, the departmental capital, commercial, and financial center of Intibucá, 11 percent of businesses are financed by remittances of returned migrants. Many of the returning migrants return at some point to the United States for temporary work and leave a family member in charge of their business.

Honduran migrants in the United States tend to cluster in areas with a high presence of peers from their home communities. Migrants from Intibucá tend to group in the Greater Washington, DC Metro Area, migrants from Olancho in Miami, and the Garífuna traditionally settle in the Bronx or Brooklyn, New York. Complementary to their transnational networks, migrants establish specific subnational remittance corridors. A subnational remittances corridor and their related transnational networks of migrants and families create a transnational bridge characterized by people, goods, money, and information moving/travelling back and forth between the place of a migrant's origin and the destination. The concept of a transnational bridge—bringing together senders and beneficiaries of the same origin—was the marketing strategy of one financial institution to promote their products and services through social corporate investment in education. Honduras faces an opportune time to strengthen transnational bridges with the assistance of many stakeholders. Only recently, financial institutions have started to look for ways to reach out to both senders and beneficiaries of remittances in order to cross-sell financial products and promote financial inclusion.

Lessons from case studies of transnational bridges suggest that understanding subnational remittance corridors and their underlying transnational migrant networks help design and implement more efficient outreach and financial inclusion even if on a low startup scale. A subnational perspective helps turn informal migration patterns to local development opportunities, builds trust, and engages key stakeholders at a local level.

Policy Recommendations

As an outcome of the analysis in this report, key policy recommendations suggest actions for stakeholders with respect to the U.S.-Honduras market for remittances, strategies for financial inclusions of senders and recipients, and development impact of remittances in rural Honduras.

The U.S.-Honduras Market for Remittances

Develop distribution channels in rural areas. The development of a payment system infrastructure can facilitate efficient transactions, including remittances that then lead to reduced costs of payment transactions. Better access can ease remittance distribution on several levels in remote areas. First, it addresses security and cost issues by avoiding carrying cash to remote payment outlets in armored vehicles. Second, the private sector entities can utilize payment infrastructure to develop fast, inexpensive, and secured remittance products that meet users' needs. Third, flexibility in access to certain payment systems, such as payment cards, by all new, authorized operators could facilitate further distribution of remittances. The Central Bank should continue to lead this effort.

Clarify regulatory requirements and compliance. KYC requirements in Honduras appear unclear for the private sector. The CNBS and UIF and other authorities should clarify, in particular, the need for physical presence of a customer at the time of opening a bank account, among other requirements. This ambiguity allows migrants working in the United States to open accounts in Honduras without being present. The quality of enforcement of KYC requirements done by these banks is unknown. The CNBS, UIF, and other authorities should take a balanced approach between the mitigation of AML risks and the improvement of access to financial services.

Regulate money transfer companies. The CNBS has drafted regulations for money transfer companies. The Honduran government authorities in collaboration with financial institutions should implement new regulations in a gradual manner in terms of requirements and timing. The regulatory framework should be sound, predictable, non-discriminatory, and proportionate. It should address transparency, ensure consumer protection, and require money transfer service providers to be held accountable for their services. Too complex requirements at the beginning for those newly regulated may discourage them from being licensed and operating legally.

Develop a monitoring/supervisory framework. The Honduran government authorities in collaboration with financial institutions should consider developing a money-laundering risk identification framework that monitors geographic risks, increased security concerns, and smuggling issues. The application of risk factors in monitoring and supervision will facilitate its effectiveness and better use of financial and human resources. The UIF is well positioned to develop a risk identification framework.

Form a committee for data collection. Currently different entities of the authorities collect remittance and migration data. The government of Honduras could consider forming a national committee to maximize available resources for better data collection. The committee could bring key stakeholders including INE, the Central Bank, the CNBS, the UIF, the Ministry of Foreign Affairs, and others together to exchange information on data and to produce better information through coordination.

Better harmonize and coordinate state regulations and examinations of money service businesses in the United States. While U.S. state regulators have voluntarily made efforts to harmonize state regulations for money service businesses, there are gaps in requirements and procedures for licensing money service businesses, which result in higher costs for business operation. State regulators should continue to harmonize regulatory requirements for these licenses. Examinations of money service businesses by state regulators and the U.S. Internal Revenue Service should be better coordinated to focus on the examinations of high-risk money service businesses.

Strategies for Financial Inclusions of Senders and Recipients

Promote inclusion and expand access with proper identification. Currently, U.S. authorities do not take positions on use of consular identification cards by undocumented migrants.[3] Many commercial banks in the United States accept consular identification cards as a form of identification for migrants. In order for Honduran migrants to enjoy this privilege, the Honduran authorities in collaboration with U.S. government authorities should develop capacity to issue secured consular identification cards for Honduran migrants in the United States.

Raise awareness of need for financial education. Honduran consulates, financial institutions, and migrant communities should work with ongoing efforts by regional offices of the U.S. Federal Deposit Insurance Corporation to raise awareness and conduct basic financial education among Honduran migrant communities.

Improve capacity of the public sector. In order to implement the above policy recommendations, the Honduran authorities should improve the capacity of their consulates in the United States for issuing secured national identification cards and consular identification cards. This will help undocumented migrant workers gain access to financial services, if these cards are considered secured by financial institutions. The Honduran government authorities should enhance the capacity of Honduran consulates to serve the large Honduran migrant population in the United States in other areas of need.

Development Impact of Remittances in Rural Honduras

Create matching fund programs for migrant's community investments. Other countries in the LAC Region have created public or private matching fund programs that complement migrants' investment in their home communities' social infrastructure. Migrants associations usually register with their consulates and compete for extra funding through their project proposals. Beyond the positive effect of additional social infrastructure in migrant's home communities these programs help to connect migrant associations to initiatives of local development and can ultimately develop partnerships for dialogue exchange.

Create migrant friendly investment policies at the local level. Some migrants plan to go back to their hometowns and invest their savings to create an income for themselves and their family. Others might be interested in helping a family member with their business idea. Local development agencies, municipalities, or others could help these migrants develop investment ideas and business plans by providing information on topics such as the following: the local economy (prices, competition, lack of products or services, investment opportunities, and so forth) business courses, legal and fiscal requirements, and sources of additional financing. Additionally fiscal incentives could be an adequate measure to attract migrant's investments back home.

Strengthen export of nostalgic products. Migrants in the United States create demand for locally produced goods, especially foodstuffs and other typical items, which often cannot be bought abroad or, when available, do not taste the same. Local goods create a nostalgic bond with the hometown. The demand for locally produced goods presents a new and growing market for local producers who often already send their products to the United States through viajeros. Formalizing and amplifying these exports are challenges. For example, local producers might need help in getting sanitary registration, export licenses, information on necessary permits and transport, and how to commercialize their products in places where migrants live.

Connect and incorporate talent abroad. Connecting highly skilled migrants to development of local-level initiatives creates opportunities for knowledge transfer and innovation. Identifying talent and creating networks of these intrinsically motivated people is a strategy applied by some countries to connect their business and scientific communities to top-level knowledge and provide them with contacts; other connecting strategies are mentoring or internship programs.

Notes

[1] Ratha and others 2008.
[2] World Bank 2009.
[3] Mexican and Guatemalan consulates in the United States issue their consular identification cards to their own nationals.

Acronyms and Abbreviations

AML	Anti-money laundering
ATM	Automated teller machine
BCH	*Banco Central de Honduras*
BSA	Bank Secrecy Act (U.S.)
CAMR	*Centro de Atención al Migrante Retornado*
CFT	Combating the financing of terrorism
CNBS	*Comisión Nacional de Bancos y Seguros*
CPSS	Committee on Payment and Settlement Systems
CTR	Currency Transaction Report
DEA	Drug Enforcement Agency (U.S.)
DHS	Department of Homeland Security (U.S.)
FBI	Federal Bureau of Investigation (U.S.)
FDIC	Federal Deposit Insurance Corporation (U.S.)
FI	Financial institution
FONAMIH	*Foro Nacional para las Migraciones en Honduras*
FRB	Federal Reserve Board (U.S.)
GAO	Government Accountability Office (U.S.)
GDP	Gross national product
GTZ	*Deutsche Gesellschaft für Technische Zusammenarbeit*
IADB	Inter-American Development Bank
INE	*Instituto Nacional de Estatistica*
INM	*Instituto Nacional de Migración de México*
IOM	International Organization for Migration
IRS	Internal Revenue Service (U.S.)
KYC	Know your customer
LAC	Latin America and the Caribbean region (World Bank)
MFI	Microfinance institution
MoFA	Ministry of Foreign Affairs
MSB	Money service business
MTO	Money transfer operator
NCUA	National Credit Union Association (U.S.)
OCC	Office of the Comptroller of the Currency (U.S.)
OPD	Private organizations for development
OPDF	*Organización Privada de Desarrollo Financiero* (Microfinance NGO)
OTS	Office of Thrift Supervision (U.S.)
RDS	*Red de Desarrollo Sostenible*
RSP	Remittance service provider
SAR	Suspicious Activity Report
STR	Suspicious Transaction Report
TPS	Temporary protection status
UIF	*Unidad de Información Financiera* (Financial Intelligence Unit)
USSS	United States Secret Service

CHAPTER 1

Overview of Migration and Remittance Trends

In 2008, the environment surrounding remittances dramatically changed along with the deteriorating economic situation spreading across the globe. The year began with an existing weak U.S. dollar, high oil prices, and a housing sector crisis caused by risky subprime mortgages. The U.S. financial downturn immediately spread into an international financial crisis, resulting in slowing economic growth on a global scale. Remittances were no exception to the negative impact of the financial crisis as an economic slowdown in migrant-host countries affects employment and incomes.[1] Consequently, the current financial crisis impacts negatively on remittances for Latin American countries whose incoming remittances are mainly from the United States.

The chapter provides an overview of migration and remittance trends. It reflects stages in the migration process and its financial implications in the natural sequence of migration: reasons for migration, cost of migration, economic background, regular and irregular migration to the United States, work in host country, return migration, remittance flow to Honduras, use of remittances, and sustainability of flows. This report addresses primarily economic issues that current migrants face; however, it has no intentions to promote further migration.

Rapid changes in the remittance environment have had implications in the preparation of this report. Although the authors tried to include updated information, fast-changing conditions have made this difficult to achieve. The report focuses on relevant public policy issues on remittances and related concerns such as access to finance, regulatory issues, the essence of remittance market, and community initiatives (transnational bridges). The study missions in the United States and Honduras were undertaken in April 2008.

Key Migration Trends

The United States is the primary destination for Honduran migrants. In 2006, 91.4 percent of remittance senders lived in the United States, 2.2 percent in Mexico, 2.1 percent in Spain, 1.9 percent in Central America, and 2.3 percent in other countries.[2] The five destinations in U.S. states for most Hondurans are Florida, New York, California, Texas, and New Jersey. But factors including the U.S. economic slowdown have spurred migrants, new and already established, to pursue opportunities in other states.[3]

The devastating effects of Hurricane Mitch in 1998 and the subsequent economic slowdown were the principal push factors that triggered a wave of Honduran migration. The main migration push factors for neighboring El Salvador, Guatemala, and Nicaragua were armed conflicts, civil wars, and counterinsurgencies. In the case of Honduras, however, international migration is much more recent; 87 percent of Hondurans migrated in the last 10 years.[4] According to the American Community Survey (ACS) of 2007, the U.S. Census 2000, and research by the Mumford Center, Honduras experienced the most rapid growth in terms of migrants from 1990–2000 of all Latin American countries (Figure 1.1). Before 1990, 24.7 percent of all Hondurans migrants entered the United States. Between 1990 and 2000, there was the highest growth rate (34.3 percent) of immigrants to the United States; and after 2000, 40.9 percent left their native country. While Instituto Nacional de Estatística (INE) statistics also reflect migrants on an irregular status, the ACS could have underreported undocumented migrants.

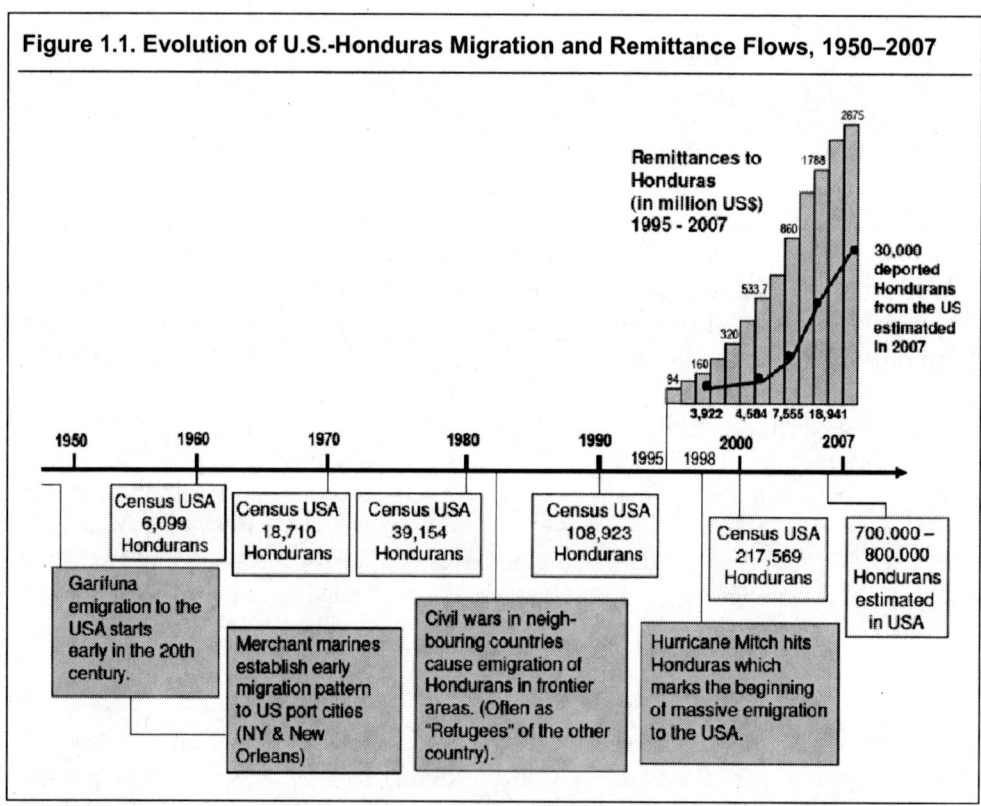

Figure 1.1. Evolution of U.S.-Honduras Migration and Remittance Flows, 1950–2007

Sources: Graph elaborated by authors based on information from U.S. Census, BCH, CAMR, England (2006), and PROMYPE/GTZ (2007).

Principle Migration Factors

According to an INE study, 91 percent of Honduran migrants emigrated in search of jobs. But migration is not always triggered by the lack of job opportunities but also by income differentials and wage gaps for unskilled labor.[5] In the department of Olancho, for example, before leaving Honduras, the majority of migrants had employment. Eighty-nine percent of male emigrants had work. Among female emigrants 42 percent worked outside the household while 31 percent worked in households before migrating. Less than 1 percent of the potential migrants were looking for work.[6] Essentially, workers chose to migrate to seek better income opportunities.

Strong social networks between migrants and their relatives seem to support and facilitate Honduran migration to the United States. Thirty percent of Honduran households have a parent or close friend living abroad who is willing to help a household member to migrate, according to the INE.[7] These familial, kinship, and ethnic networks provide information (admission policies, work opportunities); assistance (housing); and financial resources (migration costs). Social networks reduce the total costs of migration and decrease its risks. These networks contribute to a continuing process of migration that has its own dynamics independent of push factors such as the demographic and economic situation in the country of origin.[8]

A majority of migrants borrow money from family members or friends when they migrate to the United States. A third of migrants borrow money from their family members in Honduras, followed by 13 percent of family members in the United States. About 15 percent of migrants receive credits from local lenders or financial institutions to finance their migration (Box 1.1). After arrival in the United States, migrants pay an estimated US$500 monthly for seven to eight months to clear their journey cost. Attractive funding arrangements make it easier to raise money for migration than for productive purposes in Honduras.

Rural areas have higher emigration rates while urban areas have the higher number of migrants in absolute terms. In relative terms (comparing households with migrants to the total of households in one department or region), the rural departments of Colón, Olancho, and Yoro have the highest outflow of international migrants in Honduras. On the other hand, with respect to the absolute number of migrants by department, large numbers of migrants have come from Cortes (21.7 percent) and Francisco Morazán (15.7 percent). These two departments are considered to be the most urban and developed region in Honduras.[9] Figure 1.2 provides a map showing the regions (departments) of Honduras.

There are more male Honduran migrants than female migrants. Among the overall Honduran population, 70.4 percent of migrants are male and 29.6 percent are female. Almost 60 percent of migrants are between 20 and 34 years old.[10]

Honduran migrants have a higher average education achievement than non-migrants. According to the ACS 2007, 75 percent of the Honduran-born population finished the high school level or lower. Before 1998, 54 percent of the migrants had achieved only a primary education or less, rising to 63 percent in 2006. In general, migrants from Central America and Mexico come with the least educational background compared to the Caribbean and South America, but Hondurans in the United States have a higher share of migrants with tertiary education than Mexicans and other Central Americans.[11]

Box 1.1. The Alternative Remittance and Migration System

Four different service functions in the informal economy of migration are delivered to senders and beneficiaries of remittances. This informal economy is coordinated in Honduras. It is expected that because of the rising number of irregular migrants, more migrants look for the services in the informal economy.

The **money lender (*prestamista*)** lends money to a potential migrant so that he can pay the **human smuggling scheme (*coyote*)**. According to interviews in Eastern Honduras the prestamistas' interest rate varies between 10 to 15 percent a month. Between 15 to 40 percent of migrants apparently look for this service. The money lender takes as collateral tradable and nontradable assets of households with international migrants.

The human smuggling scheme, personalized by the coyote or guide, brings the migrant to the country of destination. Apparently the majority of migrants who do successfully enter the United States use the help of a coyote. The price for services of a coyote rose from US$4,000 in 2006 to US$6,000 in 2008 due to stricter U.S. admission policies. According to the National Commission for Banks and Insurance (*Comisión Nacional de Bancos y Seguros* or *CNBS*) in Honduras, the principal concern of financial institutions that report suspicious remittances is the regular transfers due to remittance payments to coyotes that range between US$200 and US$500.

In the country of destination, the migrant looks for the services of a ***viajero*** for sending remittances in cash and species and for buying nostalgic products.

At home the beneficiary needs to exchange the cash remittance delivered by the viajero in local currency. For that purpose, the beneficiary uses the exchange rate of an unregistered local **money changer (*cambista*)**. The cambista changes money with a spread of 0.1 to official exchange rate. People pay the additional price because they know and trust this person, and because they obtain quick and less formal services than are provided by any financial institution. Thirty-five percent of remittances are paid in U.S. dollars (CEMLA and MIF, 2007).

The prestamista, coyote, viajero, and cambista make up the informal economy of the migration and remittances market. All four functions could be delivered by one and the same person since they all connect to a circular economy. Their advantage is their knowledge of the client since they share in general the same migration experience, are connected to the same transnational networks, and deliver personal services in a growing informal migration market.

Source: Elaborated by authors

Box 1.2. Links between Honduras' Internal and External (International) Migration

According to a study on internal migration and labor market in Honduras, internal migration is decreasing due to international migration.[a] Investigations on local level indicate the following pattern: potential international migrants in rural communities who do not have the means to finance their migration to the exterior (through remittances from family members living abroad, credit access or selling property) first move to the more dynamic regions for better paying jobs. Internal migrants, for example, from the department of Intibucá moved first to work in the maquila industry and later migrated to the exterior.

According to an investigation about migration and the maquila industry, 41 percent of the employees have a family member in the exterior while 24 percent of the maquila workers send money to their families in rural Honduras.[b] No data is available on how many maquila workers became international migrants. But the National Association of the Maquila Industry in Honduras confirmed that internal migration to maquila industry is a first step for international migration. The Association has been campaigning about the risks of migration with the slogan "Stay with Us" because too many of its member workers regularly quit their jobs to leave the country.[c]

Sources:
a. UNAT-UNFPA 2006;
b. FONAMIH 2007.
c. Interview with Head of Marketing, Association of Maquila Industry in Honduras (March 2008).

Figure 1.2. Regions of Honduras

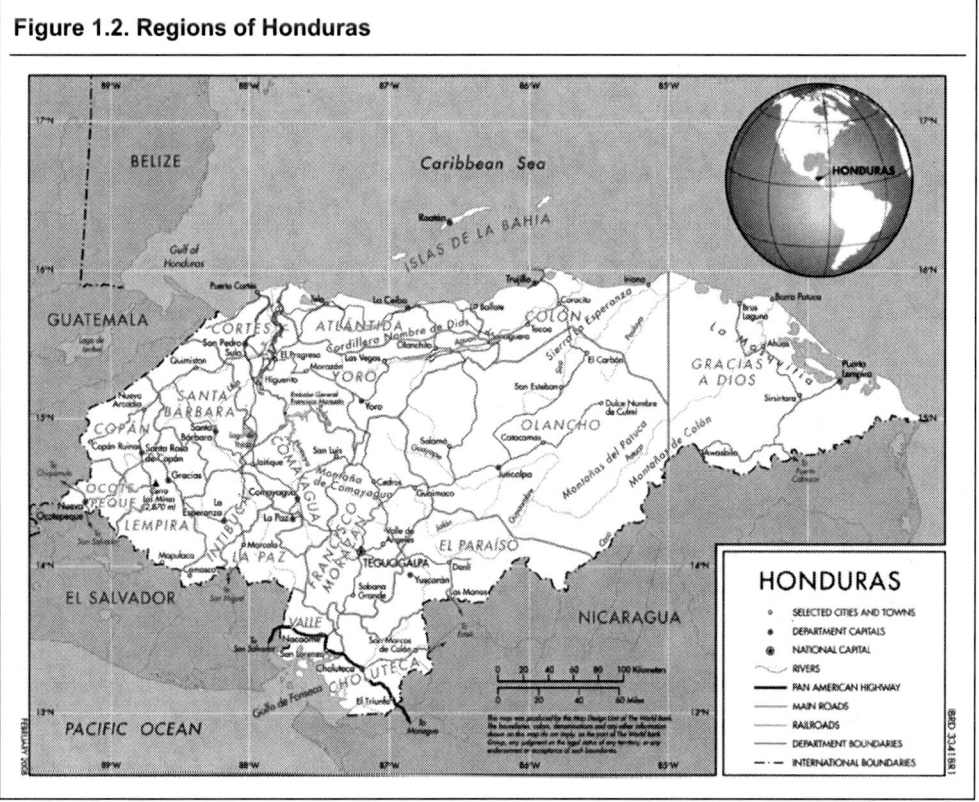

Honduran Migrants in the United States

Until recently, the growing U.S. economy attracted migrant workers from overseas, including Honduran migrants. The principle pull factor for migration to the United States had been the booming construction industry,[12] absorbing 47.9 percent of male Honduran migrant labor.[13] Hondurans are more dependent on the construction industry than any other migrant group in the United States.[14]

Estimates of Honduran migrants in the United States range widely. The 2007 ACS estimates a Honduran-born population of 430,504 in the United States.[15] The INE estimates 246,000 Honduran emigrants in the United States in 2006.[16] The Central Bank of Honduras estimated that the number of Honduran migrants in the United States is no less than 10 percent of the population, which would account for 730,000.[17] The U.S. Department of Homeland Security estimates that there are 300,000 unauthorized Hondurans in the country.[18] The rate of increase of the unauthorized Honduran population in the United States was one of the greatest along with Mexico, Brazil, India, and Guatemala. The unauthorized Honduran population increased 70 percent in the period 2000 to 2007.[19] Increasing inflow of undocumented Honduran migrants has altered the Honduran population in the United States. The portion of residents or nationalized citizens dropped from 34 percent before 1998 to 4 percent in 2006, while the portion of migrants under the temporary protected status (TPS) dropped from 32 percent in 1998 to 3 percent in 2006.[20] Temporary protection status is explained in Box 1.3.

The type of job for migrants relates to age and length of stay in the United States. According to the ACS 2007, 74.7 percent of total Honduran migrants (civilian, ages 16 or above) are the economically active migrants (labor force) in the United States; of these, approximately 70 percent are employed workers. The U.S. construction sector[21] employs 30.4 percent of Honduran labor force (47.9 percent of the Honduran male workforce); and the services occupations absorb 30.1 percent of the labor force (47 percent of the female workforce). Compared to other Central American migrants, Hondurans are the single most dependent migrant group in the U.S. construction industry, making them vulnerable to the sector's volatile dynamic as experienced in the 2008 economic crisis. Only 18 percent of Hondurans, 22 years and older, work in medium- and high-skilled labor jobs.[22] Since the majority of Honduran migrants are young people, an important issue is raised for policy makers regarding the link between migration and the incentives to obtain a better education in the United States.

According to the ACS 2007, the per capita income of a Honduran migrant in the United States is US$14,585. In comparison to Mexican, Salvadorans, and Guatemalans, a majority of Honduran migrants earn similar income, but have poorer social security and more households below the poverty threshold (23 percent), particularly among female-run migrant households (43 percent of total).[23] Only 5.8 percent of all migrants receive income with social security benefits, and 63 percent of Hondurans in the United States do not have any health insurance (the highest rate among all migrant groups in the United States).[24]

Box 1.3. Temporary Protected Status

Temporary protected status is granted to eligible nationals of designated countries. In 1990, as part of the Immigration Act of 1990, the U.S. Congress established a procedure by which the Attorney General may provide temporary protected status to immigrants in the United States when they are temporarily unable to safely return to their home country because of ongoing armed conflict, environmental disaster, or other extraordinary and temporary conditions.

The table below illustrates those countries whose nationals in the United States benefit from temporary relief from deportation.

Country	Status	Date	Numbers in 2006
Burundi	TPS	November 4, 1997–May 2, 2009	30
El Salvador	TPS	March 2, 2001–September 9, 2010	248,282
Honduras	TPS	December 30, 1998–July 5, 2010	81,875
Liberia	TPS	March 27, 1991–March 31, 2009	3,792
Nicaragua	TPS	December 30, 1998–July 5, 2010	4,309
Somalia	TPS	September 16, 1991–September 17, 2006	324
Sudan	TPS	November 4, 1997–November 2, 2007	648

Sources: Immigration Daily (www.ilw.com); U.S. Citizen and Immigration Service (www.uscis.gov).

Figure 1.3. Types of Occupations and Honduran Labor Force in the United States

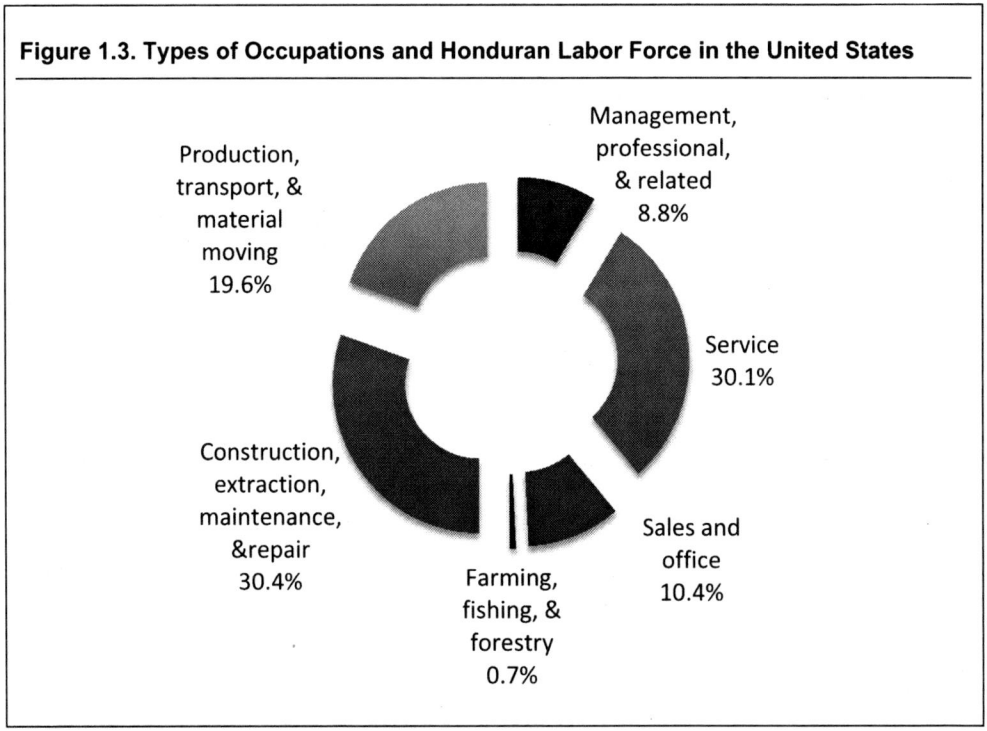

Management, professional, & related 8.8%

Production, transport, & material moving 19.6%

Service 30.1%

Construction, extraction, maintenance, &repair 30.4%

Farming, fishing, & forestry 0.7%

Sales and office 10.4%

Source: American Community Survey. 2007

Return and Circular Migration

Return migration represents an important source for local development in Honduras, yet the rising number of deportations overshadows its active promotion. Local studies and financial institutions confirm that returned migrants to Honduras are an important source of investment and employment for local business. They represent, for example, 11 percent of local business in the town of La Esperanza, the capital of the Department of Intibucá, and 6 percent of business in the rural area of Olancho. Many migrants who return to Honduras do this on a temporary basis and initiate a migration cycle.[25] Although the development impact of return and circular migration on society and economy in Honduras still requires a better understanding, three main patterns of return migration have been observed—voluntary temporary and circular migration, temporary labor programs, and deportation.[26]

Honduran migrants do not necessarily intend to stay permanently in the United States. Many migrants plan to return home or make frequent return trips to the United States thus initiating a circular migration scheme (Table 1.1). Most recent estimates indicate that the numbers of temporary migrants going to the United States have increased by an average of 10.4 percent annually. Of the 10.6 million foreign born who immigrated to the United States between 1990 and 2000, about 2.3 million returned home.[27] According to a regional study in Eastern Honduras, 70 percent of migrants return on a voluntary basis, while 30 percent are deported. Also, 8 out of 10 migrants are planning to return to the United States. Two-thirds of returned migrants are head of a household, the majority male. Thirty-one percent said they returned when they achieved a legal migration status, 28 percent said that they returned when they saved enough money, and 12 percent return regularly for tourism.[28]

Tourism is an important sector for migrants travelling back home on a regular basis and a catalyst for local economic development. Twelve percent of Hondurans living abroad travel home at least once a year.[29] Hometown tourism by migrants is developing into an important sector of the economy in some places and presents another opportunity for local economic development. Hondurans travelling back over Christmas or other holidays spend an average of US$2,273 during their stay in their home country.[30] It is important to note that this aspect of circular migration is limited to documented migrants.

Table 1.1. Routes to Circular Migration Policy

Migrant	The Usual Path: *Maintaining ties to countries of origin*	The Road Less Travelled: *Maintaining ties to countries of destination*
Permanent	Provision of return incentives	Removing disincentives to circulation: • Flexible residency and citizenship rights • Portable benefits • Accessible Information
Temporary	Restrictive temporary worker schemes	Flexible and open working arrangements: • More flexible contracts • Options of re-entry • Portability of visas • Building skills and entrepreneurship

Source: Aguinas and Newland 2007: 9.

Hondurans have become the second largest immigrant population apprehended and deported by U.S. authorities since 2000. Over 80,000 Hondurans were deported in 2006 from the United States and Mexico, and close to 70,000 in 2007. Of every 100 Hondurans who leave for the United States it is calculated that 7 percent enter regularly, 17 percent irregularly, 75 percent are deported from Mexico or the United States, and 1 percent stays in Mexico or Guatemala.[31] Forty-one percent of the deported migrants said they would leave Honduras several times until they reach their destination. The social and labor reintegration of deported migrants is a rising challenge for Honduras.

The administration of international migration through temporary labor programs is high on the agenda of policy makers, but initiatives in Honduras are still on a small scale. The Honduran government initiated two pilot approaches in 2007/08 to facilitate the hiring of Honduran workers for temporary labor contracts. One pilot is with two Canadian business associations in the food industry; the other is based on a bilateral agreement between the Honduran and Spanish governments. The long-term objective is to facilitate the annual hiring of 4,000 workers. The experience of the Guatemalan government with their IOM-assisted temporary worker programs serves as a reference. A temporary worker program on a much larger scale has been negotiated between the National Industrial Association in Honduras and a business federation in California, indicating a diversification of stakeholders in circular migration schemes.

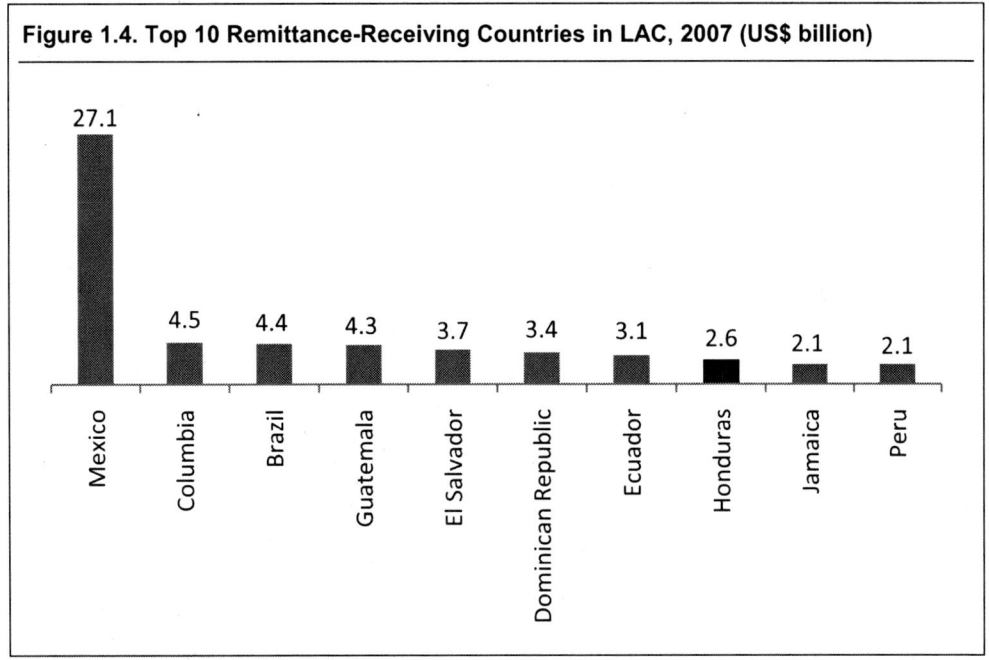

Figure 1.4. Top 10 Remittance-Receiving Countries in LAC, 2007 (US$ billion)

Source: World Bank Prospects Group (2009).
Note: Remittance inflows to Honduras in 2008 were US$2.8 billion (BCH).

Overview of Remittance Flows

Honduras is a relatively large remittance-receiving country in the LAC Region. In 2008, in absolute volume, Honduras received US$2.8 billion in remittances, which makes the country the eighth largest recipient in the LAC Region, following Ecuador with US$3.2 billion.[32] The remittance inflows to Honduras account for 4.3 percent of those to the LAC Region (Figure 1.4).

In 2007, remittances to Honduras accounted for 21.3 percent of its GDP. Measured by the remittance/GDP ratio, Honduras was a top 10 remittance-receiving country in the world and the second largest in the LAC Region.[33] Other statistics on international financial flows to Honduras support that its economy is dependent on remittances. The remittances/export of goods ratio was 46.5 percent, and the remittance/foreign direct investment ratio was 319.1 percent in 2007.

Table 1.2. Key Remittance Ratio (2007)

	Percentage
Remittance/GDP	21.3
Remittance/Export of goods	46.5
Remittance/Foreign direct investment	319.1
Remittance/Official development assistance	560.1

Source: World Development Indicator Database 2009 (World Bank) and the Central Bank of Honduras.

Most overseas remittances to Honduras originate in the United States. According to the INE, 91.4 percent of remittance senders in 2006 were in the United States. Within the United States, the remittance senders are concentrated in New York, Florida, Louisiana, Texas, and the Washington, DC metro area (Virginia, Maryland, and Washington, DC).

Over one-fifth of all households (20.9 percent or total of 330,938) in Honduras receive remittances.[34] More than half (55.6 percent) of these households are located in urban areas and 44.4 percent in rural areas. Of the households that receive remittances, 53.9 percent are led by a male and 46.1 percent are led by a female. This is a relatively higher amount of female-led households when compared to the total number of households where 75.3 percent are led by males and 24.7 percent are led by females.[35] Also when evaluating remittance receivers by gender, more women (67.2 percent) than men (32.8 percent) receive remittances in Honduras, which is conclusive with the fact that more men than women migrate.

Remittances constitute the third largest source of household income in Honduras and are largely used to finance basic living expenses. Remittances account for 11.1 percent of household income, followed by 42.4 percent of salaries, and 36.2 percent of other activities.[36] About 70 percent of remittances are used to finance basic living expenses (food, clothes, household items), and 12 percent used for medications, 9 percent for housing, 5 percent for education, and 4 percent for savings and investments (Figure 1.5).

Figure 1.5. The Use of Remittances (2006)

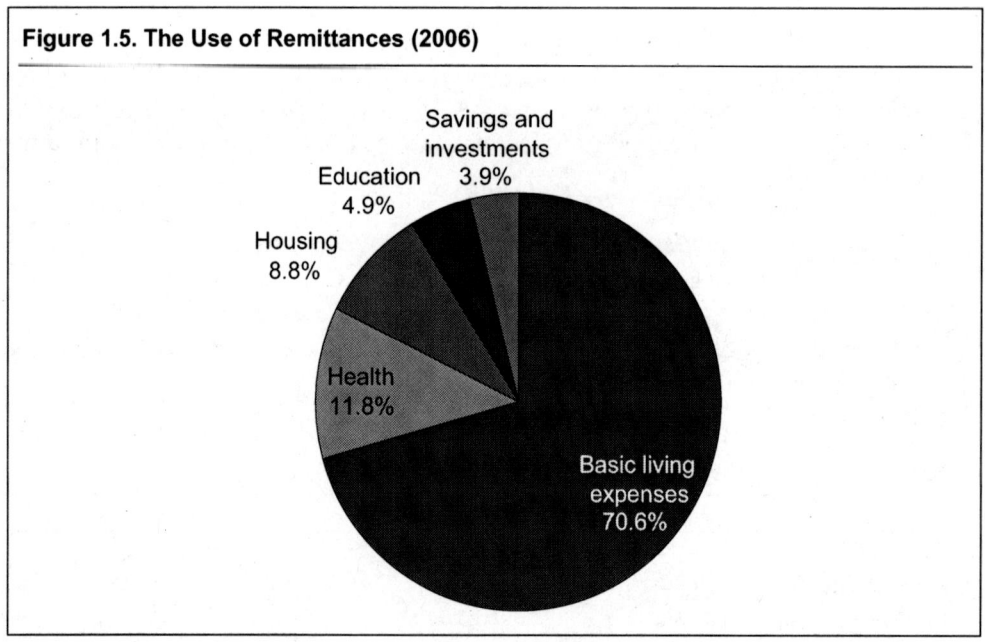

Source: CEMLA and MIF (2007).

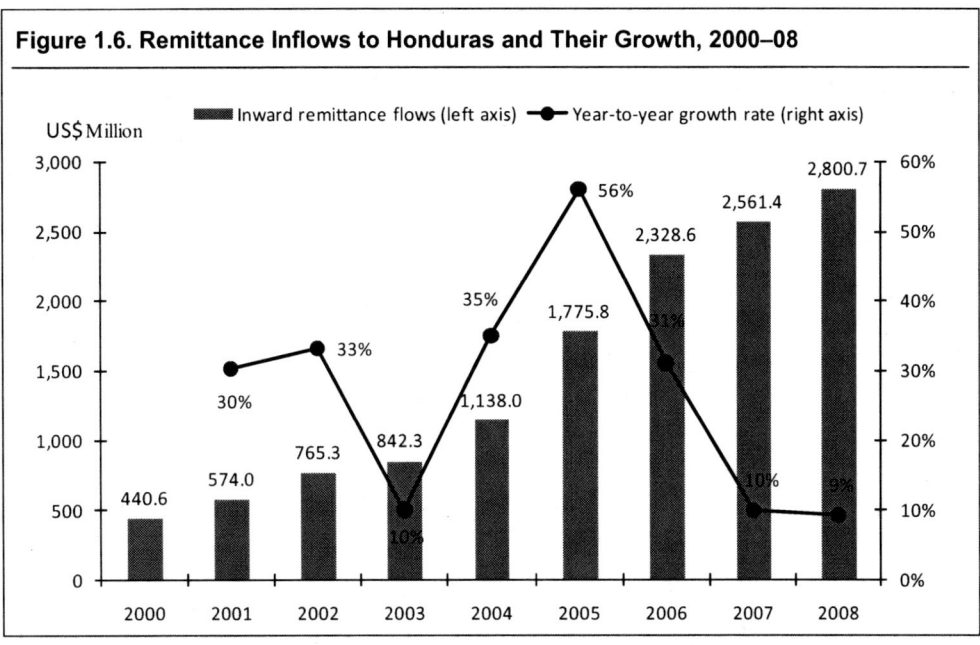

Figure 1.6. Remittance Inflows to Honduras and Their Growth, 2000–08

Source: BCH (2009).

Data on education and poverty level of remittance receivers shows that it is not only the poorest and least educated who receive remittances. According to a World Bank analysis, the average level of education in remittance-receiving households in Honduras is higher than in non-remittance-receiving households indicating a positive educational selection of migrants. Analysis of the income level of remittances recipients in Honduras shows a U-shaped distribution between income quintiles, where remittances recipients are found in the bottom and the top income quintiles, which means that the poorest and richest households receive remittances in the same proportions and more than the middle-income households. In its totality though, it appears that remittances in Honduras do lower poverty levels and inequality.[37]

Although remittances to Honduras increased to unprecedented amounts in absolute numbers, 2005–07 saw a marked slowdown in their growth rate. Since 2000, the volume of inward remittances to Honduras grew six-fold, reaching US$2.8 billion in 2008 (Figure 1.6). According to the Central Bank of Honduras, the remittance/GDP ratio has tripled between 2001 and 2006, from 8 percent. Nonetheless, since 2005, a slowdown in remittance growth rate is clearly observable. This has led to concerns about the sustainability of remittances flows and the dangers that a dependency of the economy on remittances might present.

The downward trend in the growth rate of remittances can be explained by the slowdown of the U.S. economy, tightening of U.S. and Mexican border controls, as well as data collection accuracy. The construction sector is the single most important employment sector for Honduran male migrants, but is feeling the effects of the U.S. subprime mortgage crisis and falling real estate prices. Tightening border controls between the United States and Mexico might affect the arrival of new migrants. Close to 70,000 Hondurans were deported in 2007 from the United States and from Mexico, although by far the larger part out of Mexico. Another possible factor for the downward trend in remittance growth is the adjustment for data collection. It is possible that remittances were over-reported when a new method of data collection was put in place and adjustments were subsequently made.

The Honduran government has adopted a national policy for emigrants although its implementation has been slow. According to the Honduran authorities, the government has been engaged in dialogue with neighboring states to exchange experience on migration and remittances. The Honduran Ministry of Foreign Affairs has met with policy makers in Mexico to discuss strategies on remittances and migration. Salvadorian authorities also have offered support. The Ministry of Foreign Affairs is in favor of pursuing a regional (Central American) strategy on migration and remittances, but recognizes the importance of devising first a national policy for Honduras (Box 1.4). The policy is a major attempt to build a vision of the impact of migration and remittances for Honduras, and add focus to fragmented initiatives.[38]

Box 1.4. Honduras' National Policy for Emigrants

Rising remittances entering Honduras and the ascending number of deported Hondurans arriving at the national airports and other borders have become increasing concerns in Honduran society. In 2007, the government of Honduras concluded that there was a need to create a national policy to deal with the migrant issue.

Creating policies and specialized institutions for this subject is not uncommon in the region or elsewhere in the world. Mexico created its Institute for Mexicans in the Exterior in 2003; and since 2004, El Salvador has had a Vice Minister within its Ministry for Foreign Affairs attending to the needs of its population abroad. In 2005, a National Strategy on Migration and National Action Plan on Migration were set forth by the Albanian government in cooperation with the International Organization for Migration (Albanian Government and IOM 2005).

In Honduras the process of designing the new policy was also delegated to and coordinated by the Ministry of Foreign Affairs and included consultations with other ministries and government institutions, donor agencies, civil society, as well as the private sector. The strategy included raising an inventory of existing projects and initiatives as well as defining priority areas and projects.

The idea of a National Policy for Emigrants was presented to the public in August of 2007 when Honduran President Manuel Zelaya announced the creation of a Vice Ministry for Honduran emigrants abroad, similar to the El Salvadoran experience. The final policy was presented in January of 2008.

The National Policy for Emigrants consists of three main areas:

1. **Humanitarian assistance and services** for Honduran migrants in transit countries, the countries of destination through consulate services, and upon their forceful return (integration into society and the labor market).
2. **Legalization of Honduran migration** through the negotiation of bilateral agreements for temporary labor schemes, which will allow Hondurans to work abroad temporarily and on a legal basis.
3. **Remittances and development.** Making better use of remittances for the development of the country is the main purpose of this area. Collective remittances schemes, financial inclusion, export of nostalgic products, and hometown investment of migrants are just some of the activities prioritized in this area.

The Honduran Ministry of Foreign Affairs is the coordinating agent for implementation of the National Policy for Emigrants and will therefore need to work closely with the other stakeholders and implementing agencies.

Source: Based on the National Policy for Emigrants (*Política Nacional de Atención al Emigrante)* and interviews with representatives in the Ministry of Foreign Affairs on April 7, 2008.

The government of Honduras undertook the assessment of the Committee on Payment and Settlement Systems (CPSS) and World Bank General Principles for International Remittance Services in 2007. The government has been encouraged to implement the recommendations of the assessment (Box 1.5).

Box 1.5. General Principles for International Remittance Services

The general principles are aimed at the public policy objectives of achieving safe and efficient international remittance services. Observing these principles, markets should be contestable, transparent, accessible, and sound.

Transparency and consumer protection

General Principle 1. The market for remittance services should be transparent and have adequate consumer protection.

Payment system infrastructure

General Principle 2. Improvements to payment system infrastructure that have the potential to increase the efficiency of remittance services should be encouraged.

Legal and regulatory environment

General Principle 3. Remittance services should be supported by a sound, predictable, non-discriminatory, and proportionate legal and regulatory framework in relevant jurisdictions.

Market structure and competition

General Principle 4. Competitive market conditions, including appropriate access to domestic payments infrastructures, should be fostered in the remittance industry.

Governance and risk management

General Principle 5. Appropriate governance and risk management practices should support remittance services.

Roles of remittance service providers and public authorities

A. Remittance service providers should participate actively in the implementation of the General Principles.

B. Public authorities should evaluate what action to take to achieve the public policy objectives through implementation of the General Principles.

Source: World Bank and CPSS (2007).

Notes

[1] Ratha and others (2008).

[2] INE (2007)

[3] Massey (2008).

[4] INE (2007:22).

[5] Information on wage gap for unskilled rural labor is based on interviews by PROMYPE/GTZ on Transnational Bridges in Honduras and United States.

[6] RDS and IDRC (2007b:55) Data are for the department of Olancho, Honduras.

[7] INE (2007:17)

[8] Massey (2008).

[9] INE (2007:20).

[10] INE (2007:26;27)

[11] Faijnzylber and López (2007); ACS (2007).

[12] This includes construction, extraction, maintenance, and repair occupations

[13] 2007 ACS.

[14] A comparison of construction permits issued in the United States with the flow of remittances to Honduras shows the intrinsic relation between this sector and the flow of remittances.

[15] This figure has a margin of error of 19,742.

[16] INE (2007: 23).

[17] BCH (2007a: 5).

[18] The unauthorized resident population is the remainder or "residual" after estimates of the legally resident foreign-born population—legal permanent residents, asylees, refugees, and nonimmigrants—are subtracted from estimates of the total foreign-born population. There are limitations in the data including assusmptions about undercount of foreign-born population in the American Community Survey and rates of emigration.

[19] U.S. Department of Homeland Security (2009).

[20] BCH (2007b).

[21] This includes construction, extraction, maintenance, and repair occupations.

[22] Faijnzylber and López (2007: 67).

[23] The U.S. national average poverty rate is 10 percent. The threshold is US$10,488 for single person; and US$20,444 for a family of 4 (ACS 2007).

[24] Sixty-one percent of Guatemalan, 57 percent of Mexicans, 53 percent of Salvadorans, 42 percent of Brazilians do not have health insurance in the United States. (Camarota 2007: 19).

[25] Chapter 4 gives more details of the rising transnationalism among Honduran migrants.

[26] Annecdotal information suggests that those who are deported bring little or no money back home.

[27] Agunias and Newland (2007: 5f).

[28] RDS and IDRC (2007b: 47, 65).

[29] Orozco (2007).

[30] BCH (2008).

[31] FONAMIH (2008). In 2006 59,013 Honduran migrants were deported from Mexico and 24,643 from the United States of America. In 2007 38,166 Honduran migrants were deported from Mexico and 29,348 from the United States of America.

[32] World Bank (2008b).

[33] The other top receiving countries in 2007were Tajikistan (45.5 percent), Moldova (38.3 percent), Tonga (35.1 percent), Lesotho (28.7), Lebanon (24.4), Guyana (23.5), Jordan (22.7), Haiti (20.0) and Kyrgyz Republic (19.4 percent) (World Bank 2009b).

[34] INE (2007:31).

[35] INE (2004).

[36] BCH (2007a).

[37] Faijnzylber and López (2007: 33, 38, 94, 100).

[38] Interview with representatives of Ministry of Foreign Affairs, April 7, 2008.

CHAPTER 2

The U.S.-Honduras Market for Remittances

This chapter provides an overview of the market for remittances in the United States and Honduras. It summarizes factors determining the demand for remittance service, key providers, type of remittance services, costs, level of competition, and barriers to entry. This chapter will also summarize the impact of the most relevant regulations on the remittance market in Honduras and in United States.

Senders' Preferences and Key Market Players

From interviews with Honduran migrant communities and consulates, cost is apparently not the main factor in deciding how to send remittances for remittance senders. Their choices are influenced by socioeconomic, cultural, and institutional reasons, and by their migration status. Many migrants prefer using the alternative financial services to cash their checks (two to four pay checks every month), pay bills, and send remittances.[1] Sender choices are influenced by the following main criteria:[2]

- geographic proximity of money transfer operation (MTO) in the country of destination and availability of payment locations in the receiving country
- non-bureaucratic procedures for non-documented senders
- competitive offers and promotions, such as lotteries, Mothers Day, Christmas
- extended services hours
- communication in language of the customers
- reliability and a proven record
- quick delivery times
- outlets with check-cashing and bill-paying services.

Convenience, cost, and location seem to be the major factors in determining remittance channel.[3] It is believed that new migrants integrate with existing diaspora and, given their relatively lower level of financial literacy, imitate remittance behavior that has worked well for the community in the past. According to an investigation among MTOs in New York, certain links to destination country and immigrant community where MTO is located lead to the domination of one MTO in that specific migrant community.[4]

Remittance Transfers through Formal Intermediaries

About 92 percent of remittances in the U.S.-Honduras corridor are reportedly transferred through formal remittance service providers (RSPs), although it is highly possible that some informal transfers are under-represented. In the United States, migrants use primarily large MTO networks.[5] They are attracted by relaxed requirements for money transfers under US$3,000 and do not need to provide any form of identification unless the circumstances are deemed suspicious. Retail distribution of remittances in Honduras is made possible through 16 banks, 23 credit unions,[6] 8 foreign exchange bureaus, various microfinance institutions,[7] and other commercial businesses such as supermarkets and hardware stores. Usage of banks and exchange houses are followed in popularity by MTOs (Western Union, MoneyGram, and Vigo); credit unions;[8] *sociedades financieras*; and microfinance institutions, which are referred to in Honduras as *Organizaciones Privadas de Desarrollo Financiero (OPDF)* or *Organizaciones Privadas de Desarrollo (OPD).*[9] The role of postal services in distributing remittances in Honduras is very limited. The remaining remittances are channeled to Honduras informally by viajeros and courier companies (Figure 2.1).

Market share of remittances channeled by institutions not registered to deal with foreign exchange can only be estimated. Information about market shares of players is somehow distorted because remittance flows are recorded in Honduras based on foreign exchange transactions; only commercial banks and foreign exchange bureaus are authorized by BCH as official dealers in foreign exchange, and therefore they officially account for 89 and 11 percent of the remittance market, respectively. Remittance service providers that are not authorized dealers of foreign exchange, such as credit unions, microfinance institutions, and MTOs, channel remittance transactions through an authorized financial institution.

Figure 2.1. Market Share of Remittance-Paying Service Providers in Honduras (2002–07)

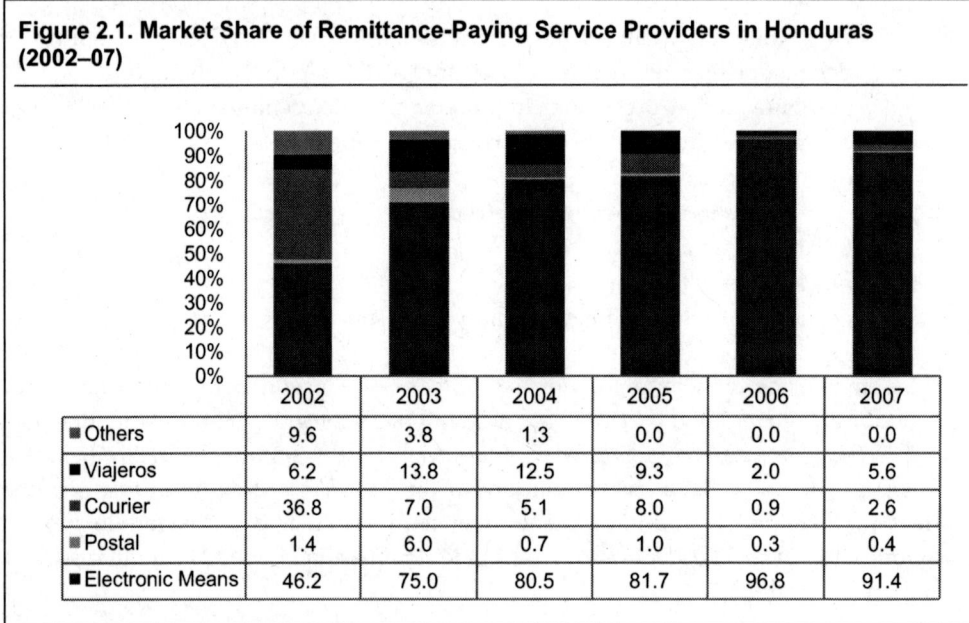

	2002	2003	2004	2005	2006	2007
■ Others	9.6	3.8	1.3	0.0	0.0	0.0
■ Viajeros	6.2	13.8	12.5	9.3	2.0	5.6
■ Courier	36.8	7.0	5.1	8.0	0.9	2.6
■ Postal	1.4	6.0	0.7	1.0	0.3	0.4
■ Electronic Means	46.2	75.0	80.5	81.7	96.8	91.4

Source: CEMLA and MIF (2007) based on data provided by BCH.

Table 2.1. Market Share of Largest Banks in the Remittance Market

	2004 (%)	2005 (%)	2006 (%)
Banco de Occidente	32.5	31.8	32.1
Banco Atlántida	18.4	14.6	18.8
Ficohsa	9.1	10.9	11.2
Grupo el Ahorro Hondureno [a]	8.2	10.7	10.8
Bamer [b]	11.0	11.6	10.1
Banpais	13.8	11.8	9.9
Other banks	7.0	8.6	7.1

Source: IDB and MIF (2007), based on data provided by BCH.
a. Now HSBC.
b. Before merger with *BAC Credomatic*.

Official statistics suggest that the remittance market is highly concentrated among few banks. Five banks hold 93 percent of all bank remittance market share, of which two hold about 50 percent (Table 2.1). In 2006, *Banco de Occidente* held about 32 percent of the market; *Banco Atlantida*, 19 percent; *Ficohsa* and *Grupo el Ahorro Hondureno*, about 11 percent each; and *Bamer* and *Banpais*, 10 percent. The other 11 banks held 7 percent of the market. The market among exchange houses is similarly concentrated. Among the eight foreign exchange bureaus that channel about 10 percent of remittances entering the country, three providers control 94 percent of this sector.

In 2005, the remittance market in Honduras was expanded to microfinance institutions (OPDFs and OPDs). Although the microfinance institution market share is still small, they are finding their niche (Box 2.1). Their growth can be attributed to their local positioning and reputation with clients and communities and due to special services at their branches (credits, insurance, and savings) and special home delivery service.

Box 2.1. The Case of Organización de Desarrollo Empresarial Femenino

With headquarters in the industrial capital of San Pedro Sula, the *Organización de Desarrollo Empresarial Femenino* (ODEF) offers specialized microfinance services to mostly female clients in northern and western Honduras. The ODEF has gone through a process of increasing supervision in the last couple of years, transforming into an OPDF in 2005 and just recently into a *sociedad financiera*. This increased supervision will allow the ODEF to grow as an institution and offer new services, such as savings accounts to non-clients.

A study on remittances in 2003 led to the conclusion that 52 percent of ODEF clients were recipients of remittances. This motivated ODEF to start an investigation into the possibilities of offering remittances as an additional client service. They were looking to reduce transfer costs, to foster savings among remittance receivers, and to attract new clients through this additional service.

The ODEF reported the following transactions in 2005–07:

Year	Transactions	Amounts (US$)
2005	2,325	1,262,219.00
2006	5,360	2,178,809.00
2007	8,430	2,416,211.00
Until June 2008	3,573	985,538.00

Source: ODEF.

Among international MTOs, Western Union has a proprietary network in Honduras and the others operate through correspondents. Western Union operates in Honduras through both a network of agents affiliated with other institutions (71 commercial banks, 13 cooperatives, 6 exchange houses) and a network of its own 11 branches.[10] Other MTOs act only as agents on behalf of financial institutions. Western Union's market share is 31–35 percent of total remittances entering the country. Through the company's distribution network, an estimated 57 percent of remittances are channeled through Western Union (contracted) agents and 43 percent is paid at its own offices.[11] Western Union's partner, *Banco Occidente*, pays out an estimated 40 percent of Western Union remittances. It is important to note that in Honduras, Western Union has a principal agent, which is a company registered as a regular business, not a financial institution. Because of a lack of regulations on the remittance businesses in Honduras, the company can provide remittance services. The company is not supervised by CNBS, although it is regulated by UIF under the AML/CFT framework.

Exclusivity contracts between financial institutions and MTOs have started to disappear. They have stopped posing a direct obstacle to competition, but their legacy explains the remaining high market concentration. Five financial institutions still have exclusive contracts with only one MTO, and the volume channeled through the largest MTOs, which had once operated with exclusivity, is still significant. There is a tendency, however, to diversify partnerships, and so most disbursing agents have started to partner with multiple MTOs. Banco Atlantida has 14 agents (MG, VIGO, Order Express, others); Banco BAC Bamer (BTS, MG, Intermex, Uniteller, Monilink, Cayman, others), and Bahncafe (Dolex, Mexico Express, MCI, Ficohsa Express, Order Express, others) also have multiple partnerships. MoneyGram partners with multiple banks and others (among them, HSBC, Banco Industrial, Atlantida, PAIZ Supermarkets); and one exchange bureau built alliances with 31 MTOs. The same is true for Western Union with contracts with Banco de Occidente, Banpais, and Ficensa.

Remittances are predominantly disbursed in cash. After cash disbursement, direct deposit is the next popular means for disbursing remittances. The usage of card-based products is not known. Exchange houses and MTOs disburse almost all remittances in cash, while banks disburse about 88 percent in cash. About 12 percent of remittances are made as direct deposit to bank accounts. The use of checks, postal instruments, and electronic checks is limited. Existing data cannot account for an accurate estimate of debit, credit, and pre-paid cash for remittance purposes.

Despite a growing network, availability of remittances services in rural areas is limited. Remittance service providers have together about 650 paying offices in Honduras (not counting the outlets of commercial businesses).[12] About 500 offices are access points offered by commercial banks, 65 by exchange houses, 90 by cooperatives, and several more by microfinance institutions. Most of the branches, as well as the network of 1,338 automated teller machines (ATMs), are concentrated in a few main cities (Table 2.2).

Table 2.2. Access Points of Remittance Services in Honduras (2007)

Institution	Offices	Branches	Access points	ATMs
Commercial banks				
Banco Atlántida, S.A.	16	87	103	101
Banco de Occidente, S.A.	8	96	104	102
Banco de Los Trabajadores	7	17	24	8
Banco Mercantil, S.A. BAMER	1	46	47	90
Banco Hondureño del Café, S.A. BANHCAFE	1	33	34	300
Banco del País, S.A. BANPAIS	2	49	51	132
Banco Financiera Comercial Hondureña, S.A. FICOHSA	1	32	33	300
Banco de América Central Honduras, S.A. BAC/CREDOMATIC	1	27	28	106
HSBC (Antes Banco Grupo el Ahorro Hondureño, S.A. BGA)	2	44	46	151
FICENSA	2	24	26	48
Total Access Points of Commercial Banks	41	455	496	1,338
Exchange houses				
Divisas Corporativas, S.A. DICORP			21	n.a.
Roble Viejo, S.A.			1	n.a.
Corporación de Inversiones Nacionales, S.A. COIN S.A.			2	n.a.
Servigiros, S.A.			41	n.a.
Total Access Points of Exchange Houses			65	n.a.
Credit and savings cooperatives				
Associated in FACACH			90	
Microfinance institutions				
Total Access Points			651	1,338

Source: CEMLA and MIF (2007) based on publicly available information.
Note: The number of ATMs includes *autobancos*, and the number of branches includes small branches with limited services.

Credit and savings cooperatives play an important role in expanding access to remittance services in rural Honduras. The cooperatives conduct 52 percent of their operations in rural areas. In 2006, cooperatives distributed about 20 percent of all remittances sent to rural areas. Cooperatives' presence in rural areas and their practice of maintaining business relationships with receivers and senders positions them well to provide other types of financial services. At the same time, they face some challenges in accessing foreign exchanges because it is confined to a channel through their own banks or foreign exchange bureax, which are authorized dealers.

Security issues have been mentioned on several occasions as a major limitation for further expanding the network of remittance-paying agents in rural areas. Transporting cash to rural and distant places is very costly because it implies making use of secure money transport firms. Banks have a comparative advantage in this aspect since they already distribute cash between their large branch networks. Conversely, each credit and savings cooperative (and sometimes microfinance institution) usually has a smaller and local branch network and holds its cash in local bank accounts (Table 2.3). Also, individual MTO agents can face difficulties by not having enough cash to pay out remittances at all times. Apparently banks transfer part of their transport costs on to other financial institutions that hold accounts with them. One credit and savings cooperative in a rural area, for example, mentioned that it pays 5 lempira for every 1,000 lempira of cash it withdraws from its bank account to be able to pay out remittances to its customers.

Remittance Transfers through Informal Intermediaries

An estimated 156 million lempira, or 6 percent of annual remittances transfers, are channeled through informal intermediaries, according to BCH. Among informal transfers, BCH distinguishes among *bolsillo* transfers (transfer of cash in bags) through friends and family members, couriers, and viajeros.

Viajeros have often been used for cash transfer services according to regional field research in Western and Eastern Honduras, despite diminishing importance. Local-level research in rural areas indicates that the importance and magnitude of informal intermediaries have been stable. The advantages of viajero remittance services are door-to-door service, willingness also to ship goods, lower charges for remittances (about 4 to 5 percent per transfer), less bureaucratic process, and payments in U.S. dollars if client requires. Competition and better, faster services among financial institutions however are thought to be the reasons for the declining significance of the viajero.

Often a viajero pools cash remittances from migrants and channels them from his U.S. bank account to Honduras. This system works best when the viajero is visiting the United States instead of carrying cash with him on the journey back to Honduras. Thus, part of these activities may be accountable in official balance of payment statistics and may in fact be greater than 156 million lempira. Usually, a viajero serves a specific transnational bridge (corridor) between a certain region in Honduras and a certain region in the United States. The business of viajeros and transnational bridges are discussed in Chapter 4.

Table 2.3. Profile of Selected Remittance Service Providers in the Honduran Remittance Market

Financial institution	Outreach Honduras	U.S. presence	MTO	Specialized financial products for senders and/or receivers	Technology/ innovation connected with remittances	Advertisement, promotion, lotteries connected with remittances	Estimated market share
Banco Atlántida	Very high	No	Multiple	Yes	Yes	Yes	High
Banco de Occidente	High	No	Multiple	No	No	No	Very High
BAC Bamer	High	No	Multiple	Yes	Yes	Yes	Medium
Banco Ficohsa	Medium	Yes	Multiple	Yes	Yes	Yes	Medium
Credit and Savings Coop. (UNIRED)	Very high	No	Multiple	Yes	No	Yes	Medium
ODEF	Low	No	Multiple	Yes	No	n.a.	Low

Source: Authors, based on publicly available information.

Table 2.4. Cost of Sending US$200 Remittance from the United States to Honduras and Other LAC Countries (percent)

	2001	2002	2003	2004 Jan	2004 Nov	2005 Dec	2008 Jan	2009 Jan/ Feb	2001–09 (%)
Jamaica	9.8	10	12.7	10.2	8.8	8.2	7.2	6.7	−31.6
Haiti	9	8.1	10.4	8.9	7.9	6.7	6.2	7.2	−20.0
Mexico	8.8	9.3	7.5	7.5	6.2	6	5.8	6.8	−22.7
Honduras		6.9	6.9	7.2	6.2	5.8	4.7	6.0	−13.0
Guatemala	7.4	7.3	7.8	7.1	6.3	5.6	6.6	5.8	−21.6
Nicaragua	7.5	7.5	7	6.9	6.7	5.2	n.a	n.a.	−30.7 *
El Salvador	6.7	6.2	5.8	5.7	5	5.2	4.6	4.1	−38.8

Source: Orozco (2006), World Bank (2008/2009).
* % shows the decline from 2001 to 2005.

Cost of Remittance Transfers

Costs of sending remittances to Honduras are low but not the lowest in the corridors from the United States to Latin America. By January 2009, the transaction cost to send US$200 to various countries in Latin America had dropped significantly in the past decade (Table 2.4).[13] According to the World Bank's Remittance Prices Database, remittances from the United States to Ecuador dropped to below 4 percent in January 2009. In general, there are several factors that appear to contribute to these cost reductions in remittances in this corridor. Total volume of remittance flows to Latin America has been increasing until recently, and the size of the remittance market has grown. With the growing market size, more remittance service providers entered the market, which resulted in increased competition. Interestingly, however, some corridors from the United States to Latin America have seen increased costs in the past year, including the U.S.-Honduras corridor.

Total costs of sending and claiming a remittance in the U.S.-Honduras remittance corridor result primarily from commission paid by sender at origination. Commission costs are distributed among the capturing agent, the intermediaries/network, and disbursing agent. Other costs include differentials between the official and unofficial foreign exchange for remittances sent is U.S. dollars and disbursed in Honduran lempira and any other indirect costs associated with claiming remittances in Honduras. Cases of commission paid by recipient at destination have not been observed.

Despite being consistent with the regional median, costs in the U.S.-Honduras corridor show high variability in terms of commission paid by sender at origination. In order to send US$200 from the United States to Honduras, migrants pay between 1.5 to 32.5 percent of the amount depending on the remittance service provider. Commercial banks usually follow a flat-fee structure charging US$30-60 for cash to bank account, or bank account to check transfers, while MTOs charge less for electronic fund transfers. Table 2.5 illustrates types of services, fees, and delivery speed of a US$200 transfer, while Figure 2.2 presents a breakdown of fees by various transfer amounts.

Table 2.5. Remittance Cost to Send US$200 from the United States to Honduras by Remittance Service Providers

Firm Name	Firm Type	Fee (USD)	Exchange rate margin (%)	Total Cost Percentage	Total Cost USD	Transfer Speed	Network Coverage in Honduras	Date
Dolex Dollar Express	MTO	8.00	0.01	4.01	8.02	Same day	Nationwide- Banco Ficohsa, BanCafe	23-Jan-09
Giros Latino	MTO	5.00	1.93	4.43	8.86	Same day	Nationwide- Banco de Occidente, Banhcafe, Banco Atlantida, HSBC	23-Jan-09
Ficohsa Express (USD)	MTO	10.00	0.00	5.00	10.00	Less than one hour	Nationwide- Banco Ficohsa	23-Jan-09
Banco Atlantida (banco atlantida	Bank	10.00	0.01	5.01	10.02	Less than one hour	Nationwide- Banco Atlantida	23-Jan-09
Vigo	MTO	8.00	1.93	5.93	11.86	Same day	Nationwide- Facach, BAC/Bamer Elektra, Banco BGA (Giros Latinos), Banhcafe, Servigiros, Banco Atlantida, Banco Ficohsa, Banco Cuscatlan, HSBC	23-Jan-09
Multivalores	MTO	9.00	1.62	6.12	12.24	Same day	Nationwide- Banco Uno, Banco Ficohsa, Banco Atlantida, Credomatic	23-Jan-09
Ficohsa Express	MTO	9.00	2.14	6.64	13.28	Less than one hour	Nationwide- Banco Ficohsa	23-Jan-09
Money Gram	MTO	9.99	1.90	6.90	13.80	Less than one hour	Nationwide- Banco Azteca, Elektra, Banco Atlantida, Supermercados Paiz, Despensa Familiar, HSBC, Banco Ficohsa, Facah, Banco Cuscatlan, Hiperpaiz, Banco Bamer, FACACH	23-Jan-09
Viamericas	MTO	10.00	1.93	6.93	13.86	Same day	Nationwide- HSBC, Credomatic, Ficohsa, Banco Bahmer, Banco de Occidente, Banco Atlantida, Banco Cuscatlan, Servigiros	23-Jan-09
Western Union	MTO	9.99	1.94	6.93	13.86	Next day	Nationwide- Banco Procredit, Bancovelo	23-Jan-09
Western Union	MTO	11.99	1.94	7.93	15.86	Less than one hour	Nationwide- Banco Procredit, Bancovelo	23-Jan-09
MTO Average		9.10	1.53	6.08	12.16			
Bank Average		10.00	0.01	5.01	10.02			
Total Average		9.18	1.40	5.98	11.98			
Total Average in 2008		8.90	0.26	4.70	9.40			

Source: World Bank Remittance Price Database 2009.
Note: Giros Latinos also provides remittance services online, which costs US$10 to send US$200.

Figure 2.2. Remittance Cost Trend to Send $200 from the United States to Honduras by Remittance Service Provider—Comparison between May 2008 and January 2009

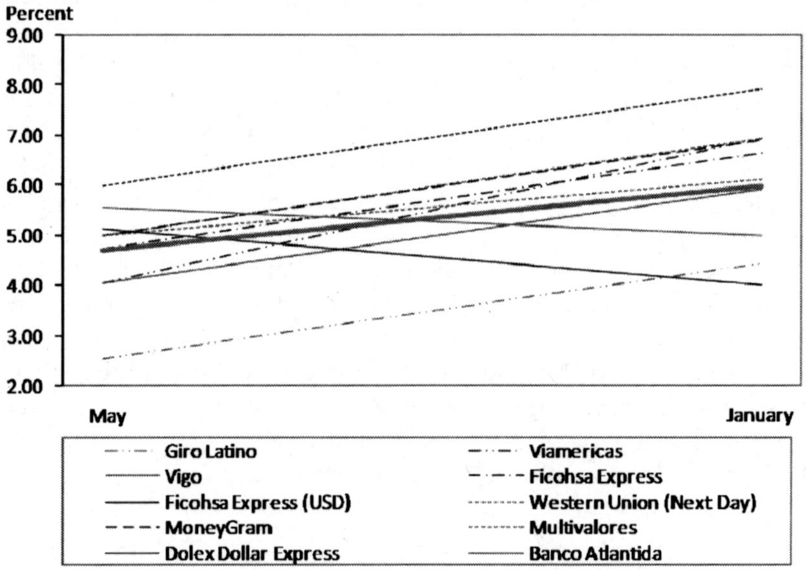

Source: World Bank Remittance Price Database 2008 and 2009.

Figure 2.3. Remittance Fees for a Range of Transfers from the United States to Honduras

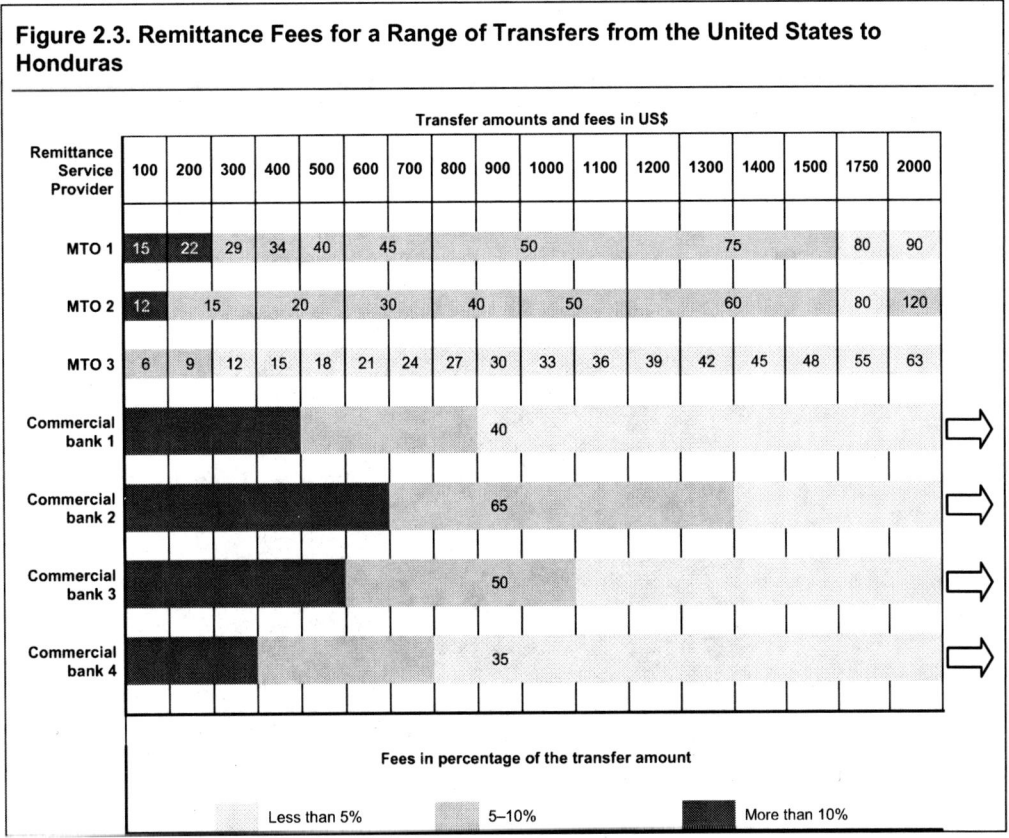

Transfer amounts and fees in US$

Remittance Service Provider	100	200	300	400	500	600	700	800	900	1000	1100	1200	1300	1400	1500	1750	2000
MTO 1	15	22	29	34	40	45		50					75			80	90
MTO 2	12	15		20		30		40		50			60			80	120
MTO 3	6	9	12	15	18	21	24	27	30	33	36	39	42	45	48	55	63
Commercial bank 1									40								
Commercial bank 2									65								
Commercial bank 3									50								
Commercial bank 4									35								

Fees in percentage of the transfer amount

Less than 5%   5–10%   More than 10%

Source: Compiled by staff based on data reported by remittance service providers.

In 2008, average cost to send $200 from the United States to Honduras began to increase. According to the World Bank Remittance Price Database, 8 out of 10 remittance service providers increased their fees. Most of these remittance service providers increased foreign exchange margins rather than fees charged at the window, which resulted in the increase of total costs of transfers. The average cost increase in the period was 20.9 percent.[14] Figure 2.3 shows remittance fees for a range of transfers in the U.S.-Honduras corridor.

Depending on partnerships and destinations, remittance service providers in the U.S.-Honduras corridor have different pricing schemes. For instance, the MTO Vigo charges US$10 for a transfer of US$1,000–1,500 for disbursing remittances on behalf of cooperatives associated in UNIRED, an acceptable fee for an average-size remittance transfer and one that allows cooperatives to compete in the market. Western Union follows the same fee structure for Honduras as for El Salvador, Guatemala, and Nicaragua. It charges slightly more for transfers to Costa Rica and less for those to Mexico. MoneyGram, on the other hand, charges the same fees for transfers to all countries in Central America and Mexico.[15]

Box 2.2. Ficohsa Express: Expansion of a Honduran MTO in the United States

Following five years of growth, Ficohsa Express operates 15 MTO branches in seven U.S. states—Florida, Georgia, Louisiana, New Jersey, New York, North Carolina, and Virginia. In 2006, the branch in Tampa, Florida observed a 32 percent increase of remittances (up to US$45 million) and a 34 percent increase in the number of transactions. The MTO estimates that about 35 percent of remittances sent to Honduras are received in cash, 35 percent to beneficiaries' bank accounts, and 30 percent to the migrant's account in Honduras. In addition to remittance services, Ficohsa Express provides other services for the migrants such as purchase of flight tickets, shipment of goods, telephone service, and bill payments.

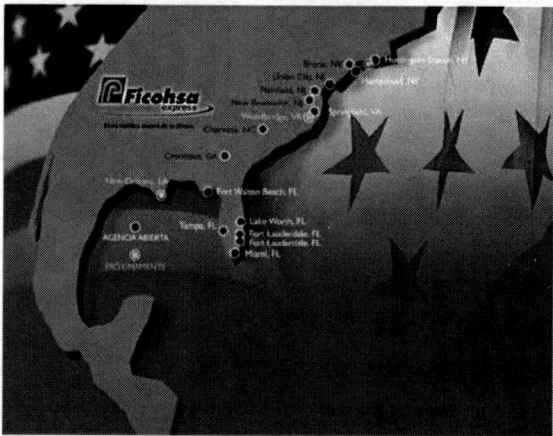

Source: Staff interviews, and Ficohsa website http://www.ficohsa.com/2007/banco/fexpress.html.

Disbursing agents in Honduras receive the least share of commission paid by sender. Depending on the arrangement, the disbursing agent charges 8 to 30 percent of commission paid by the sender. The remainder is usually divided among the capturing agent and the provider of the payment network/platform. Forty percent each for capturing agent and intermediator and 20 percent for the disbursing agent is the breakdown negotiated by one of the largest providers of remittances to Honduras. Such a breakdown of commission may explain the incentives of some disbursing agents in Honduras to integrate vertically and upwards the industry's value chain, and to set up networks of capturing agents in the United States (Box 2.2). The reverse downward trend can also be observed by foreign capturing agents' expansion in Honduras as independent disbursing agents and owners of proprietary payment platform solutions.

The actual exchange rates applied to remittance transfers sent from the United States often deviate from the official rate determined by BCH in its daily auctions. The lempira to U.S. dollar exchange rate operates under a "crawling band" regime and has stabilized in recent years. It became practice for remittance service providers to refer to the official exchange rate, thus making the interest rate cost insignificant to senders and receivers. Some service providers however do not follow this practice or are not aware of the Honduran foreign exchange regime and so adjust the foreign exchange rate arbitrarily to lempira. Since only banks and exchange houses are authorized to deal in foreign exchange, other market participants (cooperatives and microfinance institutions) may expect to pay an additional commission.

Recipients in Honduras are not charged fees to claim their remittances yet bear indirect costs associated with limited availability of remittance distribution points. Although most remittance outlets in Honduras are concentrated in large urban centers (capital cities of the country's departments), 81 percent of the receiving households are located in rural areas and medium-sized urban centers. [16] Thus the majority of beneficiaries in Honduras must travel considerable distances in order to claim remittances. One of the commercial banks, BAC Bamer, estimates that the average cost of claiming a remittance from their nationwide branch network in rural Honduras is US$6.25. [17] In the geographically largest region, Olancho, an estimated 10 percent of recipients spent eight hours or more in transit to claim a transfer; significant opportunity costs and associated expenses are involved (for example, an overnight stay in a hotel).

Data errors and temporal failures by the remittance service providers continue to increase costs to recipients. Usual errors appear in the spelling of sender's name, transaction origin, transaction code, and amount. According to a bank in Honduras, in 2005, an estimated 70 percent of recipients obtained remittances at the first attempt; in 2008 the number increased to 90 percent.[18] In Catacamas, the problem apparently had been significant in the past but now is less. Banco Atlantida in Catacamas estimates an unsuccessful transactions rate of 8 percent. This still means that 1 out of 10 recipients require two or more attempts to claim funds. Also, these statistics do not account for frequent failures of service provider payment systems, which can be slowed due to software problems and lack of electricity, or cases of poor communication between the sender and the recipient.

Impact of Regulations on Remittance Markets

Appropriate levels of regulations are necessary for the remittance market in order to ensure a level playing field, transparency, consumer protection, and the integrity of remittance flows. Protecting the integrity of remittance flows is highly important to avoid criminal activity and to protect legitimate flows of money. Consumer protection should be in place to protect workers' remittances. The Financial Action Task Force (FATF) Recommendations—in particular Customer Due Diligence and Record Keeping (Recommendation 5) and Special Recommendations VI (Money and Value Transfers) and VII (Wire Transfers)—can be applied for protecting the integrity of remittance flows. The General Principles for International Remittance Services promote a sound, predictable, non-discriminatory, and proportionate legal and regulatory framework; transparency; consumer protection; and governance of service providers (refer back to Box 1.5).

Regulatory Framework Impacting Remittance Market in the United States

In the U.S. remittance market, remittance service providers are composed of commercial banks and money service businesses (MSBs).[19] The U.S. regulations govern the remittance market at both federal and state levels. The Federal Government regulates federal-charged banks and money service businesses on AML/CFT issues, while state regulators cover the state-charted banks and money service businesses on operations. The Financial Crimes Enforcement Network (FinCEN) is the administrator of the Bank Secrecy Act (BSA). FinCEN is the U.S. financial intelligence unit, a bureau of the U.S. Department of the Treasury. There are eight federal compliance examiners of BSA (Figure 2.4).

Figure 2.4. AML/BSA Framework in the United States

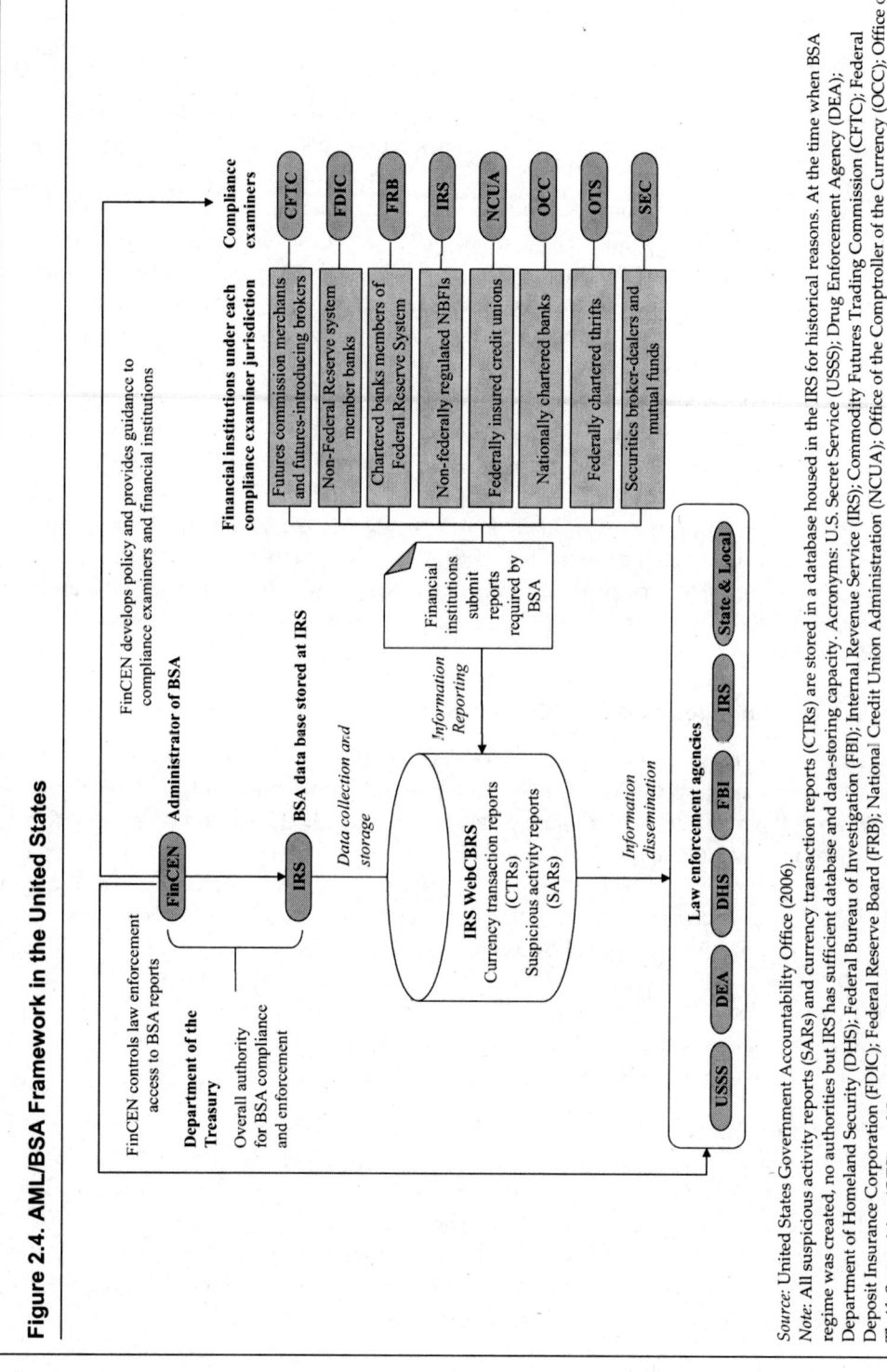

Source: United States Government Accountability Office (2006).

Note: All suspicious activity reports (SARs) and currency transaction reports (CTRs) are stored in a database housed in the IRS for historical reasons. At the time when BSA regime was created, no authorities but IRS has sufficient database and data-storing capacity. Acronyms: U.S. Secret Service (USSS); Drug Enforcement Agency (DEA); Department of Homeland Security (DHS); Federal Bureau of Investigation (FBI); Internal Revenue Service (IRS); Commodity Futures Trading Commission (CFTC); Federal Deposit Insurance Corporation (FDIC); Federal Reserve Board (FRB); National Credit Union Administration (NCUA); Office of the Comptroller of the Currency (OCC); Office of Thrift Supervision (OTS); and Securities and Exchange Commission (SEC).

Table 2.6. Status of BSA Regulations for Remittance Service Providers

Type of institution	Subject to BSA Rules	Requirements	Must have AML program?	Must file SAR?	Must file CTR?	Must file 8300s?*	Must have a CIP?
MSB	Yes	Title 31 CFR § § [103.11, 20, 22, 23, 24, 25, 27, 28, 29, 33, 37, 41, and 125]	Yes	Yes	Yes	No	No
Bank**	Yes	Title 31 CFR § § [103.11,18, 22, 23, 24, 25, 26, 27, 28, 33, 120, 177, 181, and 183]	Yes	Yes	Yes	No	Yes

Source: U.S. Money Laundering Threat Assessment.
* IRS Form 8300, *Report of Cash Payments over $10,000 Received in a Trade or Business.*
** Depository financial institutions including commercial banks, savings and loan associations (or thrifts), and credit unions.

At the federal level, money service businesses are required to comply with the Bank Secrecy Act, the U.S. PATRIOT Act, and relevant regulations and guidelines. Main requirements include registration with the U.S. Department of the Treasury and setting up an AML/CFT program, including customer identification and record-keeping and reporting to relevant authorities. [20] The main purpose of the BSA governing MSB activities is to collect reports and records of transactions that can be used for criminal, tax, and regulatory investigations or proceedings, as well as develop intelligence or counterintelligence against terrorism. Different compliance examiners and supervisors are responsible for enforcing BSA, depending on types of institutions. For the U.S.-based remittance service providers, FDIC, FRB, IRS, NCUA, and OCC are competent authorities for BSA examination (Table 2.6).

Report-filing requirements for remittance service providers at the federal level include the SAR and currency transaction report. These reports are filed with FinCEN. Banks are required to file a SAR on transactions or attempted transactions involving at least US$5,000 that the financial institution knows, suspects, or has reason to suspect any of the following:

- Involves money derived from illegal activities,
- Intended or conducted in order to hide or disguise funds or assets derived from illegal activity,
- Designated to evade BSA requirements or other financial reporting requirements (structuring), or
- Has no business or apparent lawful purpose.[21]

For money service businesses, transactions or attempted transactions involving suspicious activities or transactions involving US$2,000 or more require filing a SAR.[22] It is important to note that transactions or attempted transactions that are considered suspicious must be reported regardless of an amount. All remittance service providers (banks and money service businesses) are required to file a currency transaction report on currency transactions in excess of US$10,000.

Both banks and money service businesses are required to have an AML program. An AML program must be in writing and at minimum include (a) the development of internal policies, procedures, and controls; (b) the designation of a compliance officer; (c) an ongoing employee-training program; and (d) an independent audit function to test programs.[23] Among remittance service providers, banks are required to establish a Customer Identification Program (CIP) that implements reasonable procedures to:

- Collect identifying information about customers opening an account,
- Verify that the customers are who they say they are,
- Maintain records of the information used to verify their identity, and
- Determine whether the customer appears on any list of suspected terrorists or terrorist organizations.[24]

Money service businesses are required to verify the identify customers who send or receive $3,000 or more[25] and create and maintain the record of transactions, regardless of the method of payment.[26]

With respect to types of identification, financial institutions can accept government-issued identification at their discretion. This includes foreign-issued identification. The U.S. Treasury Department concluded that the risk-based approach taken by the final regulations implementing Section 326 of the U.S. PATRIOT Act, combined with the ability to notify financial institutions if concerns arise with specific identification documents, provide an ample mechanism to address any security concerns.[27] The remittance service providers are held accountable for the effectiveness of customer identification and verification. Some remittance service providers have accepted foreign-issued identification, such as a Mexican consular identification card. Honduran embassy and consulates in the United States do not issue such identification cards.

For state licensing, each state has different requirements in spite of ongoing efforts by regulators to harmonize state regulations.[28] Main objectives of state regulations are ensuring safety and soundness of the financial systems and protecting consumers from unfair practices. Key components of state licensing include ownership and management of a company, criminal record history, audited financial statement, an AML program compliant with federal requirements, a surety device (surety bond or a deposit), and minimum net worth maintenance. The New York Banking Department, for example, grants a license to a money transmitter after examining background reports prepared by a New York State licensed investigator,[29] management and supervisory experience in money service business, fingerprints, and financial documentation. The regulator also ensures that a money service business meets requirements under BSA/AML policies and procedures.[30]

As part of licensing processes, many states require money service businesses to submit bonds and net worth/capital in order to ensure that consumers are protected. In Florida, money service businesses are required to have US$100,000 plus US$50,000 per location or agent, up to a maximum US$500,000, as well as of minimum of US$12,500 up to maximum $250,000 for security device requirements.[31] In New York, the amount of surety bond is no less than US$500,000. This amount may be reduced at the discretion of the New York Banking Department. Money service businesses often blame these state requirements for their lower profits, especially those outfits that operate in multiple states because they feel they are pressured by the market to reduce fees while paying these dues to regulators. However, a reason behind bond requirements is to protect consumers and ensure soundness of the financial systems.

There are ongoing efforts to clarify the examinations of money service businesses, concerning which challenges may still exist in the coordination between state regulators and the IRS. State regulators conduct examinations of money service businesses, including for BSA purposes, since the IRS is also responsible for BSA examination of money service businesses, which are non-federally regulated non-bank financial institutions. The U.S. Government Accountability Office (GAO) points out that the IRS needs an effective BSA compliance program. Currently, it appears that IRS has not identified the population of non-bank financial institutions, which results in the examinations of limited non-bank financial institutions.[32] Interviews with state regulators suggest that better coordination is needed between state regulators and the IRS in order to maximize limited resources for examinations. FinCEN has taken recent steps to better harmonize the examinations of money service businesses including the publication of an MSB Examination Manual[33] in December 2008.

The U.S. government does not impose foreign exchange restrictions on remittances, except for the countries and companies subject to sanctions and embargos by the U.S. Treasury. In addition, the Office of Foreign Assets Control blocks properties and prohibits transactions with those who commit, threaten to commit, or support terrorism. As noted above, financial institutions are required to file a currency transaction report on transactions above US$10,000.

Honduran Regulatory Framework Impacting on the Remittance Market

The Law of the National Commission of Banks and Insurance (*Comisión Nacional de Bancos y Seguros* or CNBS) provides the Commission with the authority to supervise financial institutions.[34] The Superintendence of Securities and Other Institutions (*Superintendencia de Valores y Otra Instituciones*) in CNBS is responsible for the remittance market in Honduras. Until recently, a pure money transfer company was not considered to be a financial institution. In February 2008, a new law was passed to regulate remittance service providers. The CNBS has been drafting a new regulation for the law and intends to implement it after consultation with the private sector. Under the new law and regulation, remittance service providers will be required to have a license from CNBS. In the meantime, the CNBS has been making efforts to identify the scope of existing remittance market and market players.

The BCH is responsible for the oversight of national payment systems. The basic laws that govern payment and settlement systems are the BCH Law, the Financial Institutions Law, the Law on Credit Cards, and the Commerce Code. The BCH has implemented regulations that enforce these laws, including the Regulation on the Electronic Check Clearinghouse and the Regulation for the Automated Clearinghouse. Under the current legal and regulatory framework, only banks have direct access to the main payment systems because of the requirements that the users have a current account with BCH.[35]

The BCH is also responsible for foreign exchange regulations and operates a public auction system called *Sistema de Adjudicación Pública de Divisas*. Of all actors in the remittances market, only commercial banks and foreign exchange houses are authorized foreign exchange dealers. All other institutions have to carry out their foreign exchange transactions through one of these authorized agents. Banks and foreign exchange houses are required to sell the foreign exchange within three business days to BCH, which then supplies the demand for foreign exchange through its public auction system on a daily basis.

According to monetary law, financial institutions are required to pay out remittances in local currency, except for deposits in foreign currency accounts. The regulations are illustrated in Figure 2.5.

In the AML/CFT domain, the Honduran AML Law, *Decreto No. 45-2002, Ley Contra el Delito de Lavado de Activos*, passed in 2002. The law established *la Unidad de Informacion Financiera* (UIF), Honduras' financial intelligence unit. The AML Law requires supervised and other relevant financial institutions to establish formal AML policies and procedures, including appointing a Compliance Officer and Compliance Committee, KYC policies and procedures, ongoing monitoring of customers, and filing suspicious transaction reports (STRs) to the UIF. The CNBS regulation, *Reglamento para la Prevención y Detección del Uso Indebido de los Servicios y Productos Financieros* (AML Regulation), complements the AML Law.

Figure 2.5. Regulations in the Market for Remittances

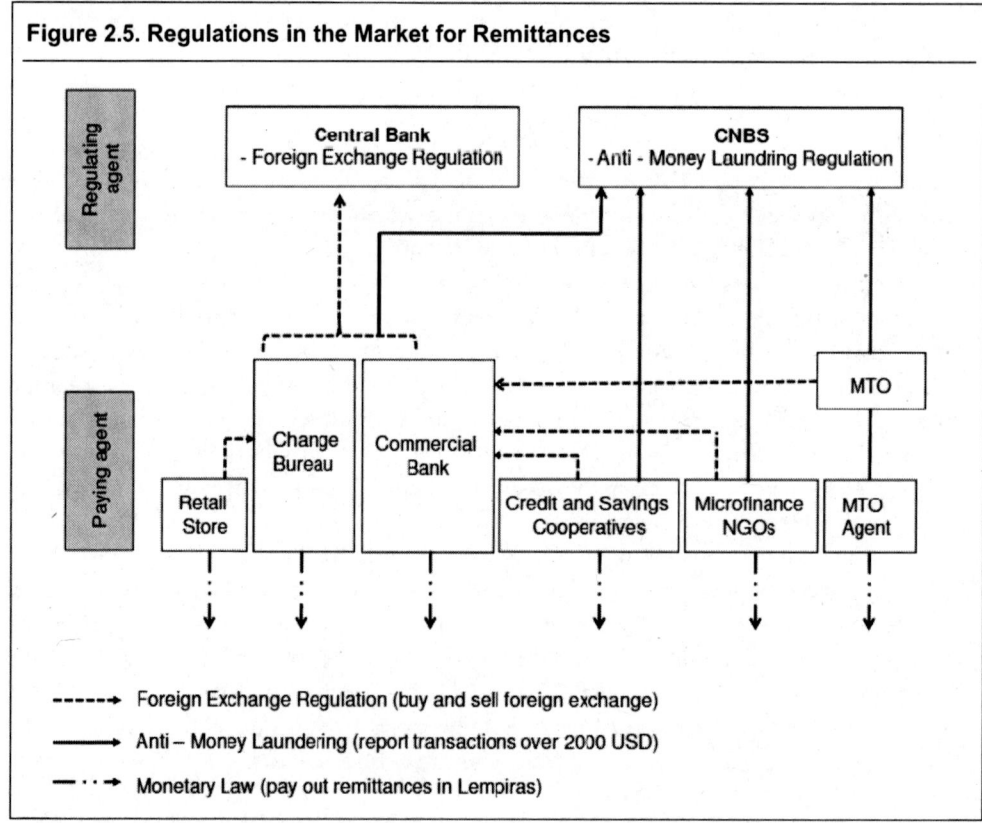

Sources: Graph elaborated by authors based on publicly available information.

The AML Regulation stipulates the requirements for customer identification. Financial institutions are required to obtain the following information from the normal clients:

- given name and family name
- identification card number
- civil status
- profession, title, or occupation
- nationality
- address
- telephone number
- company where a client works
- bank or commercial references.

Financial institutions are also required to obtain a photocopy of an identification card or a passport if a client is a non-resident foreigner.

With respect to KYC requirements for financial institutions, it seems to be unclear whether a client's physical presence is needed at the time of opening an account or making a transaction. The AML Law prescribes customer identification; however, it does not require physical presence. At the same time, the authorities interpret that physical presence is necessary, although this interpretation is not publicly issued. A few financial institutions in Honduras are opening accounts for Honduran migrants while they are in the United States.

The UIF was a pass-through of STRs to the Public Ministry before the law was amended in 2008. Since this amendment, the UIF analyzes and grades all incoming STRs and only sends those STRs to the Public Ministry which merit further action. Since the enactment of the amendment, the UIF receives reports of all transactions equivalent or larger to US$10,000 (in lempiras or other currencies) from all financial institutions subject to the AML Law.[36] Also see below reporting requirement for MTOs.

Key Findings and Policy Recommendations

Develop distribution channels in rural areas. The development of a payment systems infrastructure can facilitate efficient payment transactions including remittances. Efficient transaction will help reduce cost of payment transactions. Better access to payment infrastructures can ease remittance distribution in remote areas. First, it would address the cost of security issues by avoiding carrying cash to remote payment outlets with armed cars. Second, the private sector entities can utilize payment infrastructure to develop more attractive products that meet users' needs such as fast, inexpensive, and secured remittance products. Third, flexibility in access to certain payment systems by all new potential operators that meet relevant requirements could facilitate further distribution of remittances.[37] The Central Bank should continue to lead this effort.

Clarify regulatory requirements and compliance. KYC requirements in Honduras appear unclear for the private sector. The authorities such as CNBS and UIF should clarify the requirements, in particular, physical presence of a customer at the time of opening a bank account. Currently, this ambiguity allows bancarization of the unbanked migrants in the United States. However, the quality of KYC done by banks is unknown. The authorities should take a balanced approach between the mitigation of AML risks and the improvement of access to financial services.

Regulate money transfer companies first to create a level playing field. CNBS has prepared a draft regulation for money transfer companies. The authorities should implement new regulations in a gradual manner in terms of requirements and timing. The regulatory framework should be sound, predictable, non-discriminatory, and proportionate. It should address transparency, ensure consumer protection, and require money transfer service providers to be held accountable for their services. Too complex requirements from the beginning for those newly regulated may discourage them from being licensed and operate illegally.

Develop a monitoring/supervisory framework. The authorities should consider developing a money-laundering risk identification framework that studies geographic risks, increasing security concerns, and smuggling issues. The application of risk factors in monitoring and supervision will facilitate its effectiveness and better use of financial and human resources. The UIF is better positioned to develop a risk identification framework.

Form a committee for data collection. Currently different entities of the authorities collect remittance and migration data. The government of Honduras could consider forming a national committee to maximize available resources for better data collection. The committee could bring key stakeholders including INE, the Central Bank, the CNBS, the UIF, the Ministry of Foreign Affairs, and others together to exchange information on data and to produce better information through coordination.

Better harmonize and coordinate state regulations and examinations of money service businesses in the United States. While state regulators have voluntarily made efforts to harmonize state regulations for money service businesses, there are gaps in requirements and procedures for a MSB license, which result in higher costs for business operation. State regulators should continue to harmonize regulatory requirements for a MSB license. Examinations of money service businesses by state regulators and IRS should be better coordinated to focus on the examinations of high-risk money service businesses.

Notes

[1] Federal Reserve of Chicago (2006).
[2] Based on staff fieldwork in the United States.
[3] Based on staff interviews with market participants.
[4] Interview with Palmen Nikolov, New York State Banking Department (NYSBD), 2006.
[5] Referred to in the United States as a money service business (MSB).
[6] Twenty-two credit unions form part of the UNIRED network of *Federación de Cooperativas de Ahorro y Crédito de Honduras* (FACACH). One additional credit cooperative pays out remittances without being member of FACACH.
[7] *Oranización de Desarrollo Empresarial Feminino* (ODEF), FAMA, and ADED-Valle.
[8] FACACH estimates credit unions distribute about 3 percent of remittances. Through UNIRED network, US$49 million in 2006 and US$69 million in 2007 was paid in remittances.

[9] *OPDFs* are microfinance NGOs under Honduran banking supervision. *OPDs* are microcredit NGOs. They are not supervised and therefore they can only give credit but not take deposits.

[10] IDB and MIF (2007).

[11] BCH estimates Western Union share at 35 percent (December 11, 2007).

[12] Other sources estimate the number of available distribution points at 742.

[13] Orozco (2006); World Bank (2008b).

[14] Since there appear many factors to this increase, further research needs to be done to identify specific reasons.

[15] BCH (2007a).

[16] It is estimated that 81 percent of remittances are channeled to agents in three regions —Franciso Morazan (37 percent), Cortes (30 percent), Atlantida (16 percent)—and their largest municipalities, Tegucigalpa, San Pedro Sula and Ceiba (BCH 2007b:18).

[17] Interview with BAC-Bamer, March 2008; investigation by RDS and IDRC (2007b).

[18] Authors' interview with a bank in Honduras.

[19] MSBs are non-bank financial institutions, some of which provide cross-border remittance services.

[20] There are five types of reports: (a) Currency Transaction Reports, (b) Report of International Transportation or Monetary Instruments, (c) Report of Foreign Bank and Financial Accounts, (d) Suspicious Activity Report, and (e) Designation of Exempt Person Form.

[21] U.S. Money Laundering Threat Assessment, Appendix 3.

[22] Suspicious activity is any conducted or attempted transaction or pattern of transactions that you know, suspect, or have reason to suspect meets any of the following conditions:
- Involves money from criminal activity.
- Is designed to evade Bank Secrecy Act requirements, whether through structuring or other means.
- Appears to serve no business or other legal purpose and for which available facts provide no reasonable explanation.
- Involves use of the money services business to facilitate criminal activity (*Source*: FinCEN).

[23] U.S. PATRIOT Act and FinCEN.

[24] Factsheet by the Office of Public Affairs, U.S. Department of the Treasury, September 18, 2003, www.treas.gov/press/releases/reports/js7432.doc.

[25] 31CFR103.33(f).

[26] "FinCEN—A Money Services Business Guide."

[27] Factsheet by the Office of Public Affairs, U.S. Department of the Treasury, September 18, 2003.

[28] The Money Transmitter Regulators Association was established in part to harmonize examination standards and regulatory system. Members include 42 states and District of Columbia. For more information, go to http://www.mtraweb.org.

[29] An investigator examines all partners, officers, directors, and substantial stockholders.

[30] New York Banking Department Instruction Sheet—For a Transmitter of Money License.

[31] Bond, Certificate of Deposit, or Letter of Credit.

[32] GAO (2006).

[33] Available at http://www.fincen.gov.

[34] Financial institutions are defined in the Law of Financial System.

[35] World Bank Payment Systems Group.

[36] Resolution 325-9/2003.

[37] These types of payment systems are, for example, payment card systems.

CHAPTER 3

Strategies for Financial Inclusion of Senders and Recipients

This chapter discusses remittances as an entry point to financial inclusion of senders and receivers, and illustrates how productive use of remittances and financial inclusion respond in parallel. The chapter presents current strategies and challenges of various financial institutions—banks, credit and savings cooperatives and microfinance institutions—in bancarization of remittance senders and receivers and highlights the role of MTOs in bancarization.

Remittances and Financial Inclusion

Massive flows of remittances present a historic opportunity for Honduras to upgrade its financial sector and increase financial inclusion of the poor. Empirical studies of countries across the globe suggest that development of the financial sector and financial inclusion has a positive impact on economic growth. Financial inclusion refers to giving access to people who formerly had no access to formal financial systems and services such as accounts, credits, and insurance products. Remittances transferred through the formal financial system builds relationships with people who had no prior contact with financial institutions. Figure 3.1 charts the flow of remittances and financial development in Honduras from 1975 to 2003.

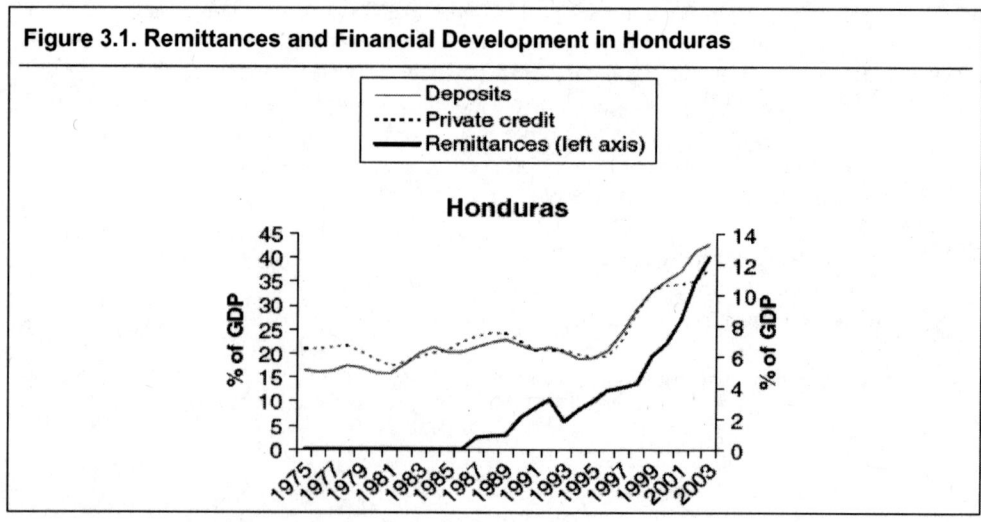

Figure 3.1. Remittances and Financial Development in Honduras

Source: Faijnzylber and López (2007).

Table 3.1. Bancarization of Remittance Recipients and Non-Recipients

	Dom. Rep.	Jamaica	Colombia	Ecuador	Bolivia
Recipient (%)	66	65	52	46	44
Non-recipient (%)	58	60	45	34	35

	Guatemala	Peru	Honduras	El Salvador	Mexico	Nicaragua
Recipient (%)	41	37	34	31	29	10
Non-recipient (%)	17	35	16	19	28	10

Source: Orozco and Fedewa (2006) based on *Receptores de Remesas en Mexico* (October, 2003); *Receptores de Remesas en Guatemala, El Salvador y Honduras* (September 2003); *Receptores de Remesas en Ecuador* (September, 2004); *Receptores de Remesas en Bolivia, Peru* (September, 2005).

Remittances to Honduras increase bancarization of remittance recipients, albeit at low overall levels. Orozco and Fedewa (2006) estimate that around 16 percent of non-recipients hold bank accounts in Honduras compared to 34 percent of remittance recipients (Table 3.1). This level of bancarization is low compared to others in the region; however, the divergence between recipients' and non-recipients' bancarization suggests that remittances have a significant impact on bancarization in Honduras, second only to Guatemala.

The potential exists for more financial inclusion of remittance recipients in Honduras. According to a survey commissioned by the Inter-American Development Bank, remittance recipients responded favorably to accessing more financial services (although their survey responses did not indicate whether they would be interested or would actually purchase financial products when offered at market price). But their interest was evident by the responses to survey questions (Table 3.2).[1]

Honduran migrants in the United States are not fully prepared to access the potentially available financial services. The amount of available information on financial inclusion of Honduran migrants in the United States is limited, but suggests that undocumented migrants cannot open accounts at U.S. banks. Most Honduran migrants do not have a valid U.S. entry visa or U.S. social security number, and often they have either lost their Honduran identification documents in transit or are too young to hold any form of Honduran identification, which leaves them with no identification documentation at all. Also, since the migrants did not have bank accounts in Honduras, they have not been sufficiently educated about the benefits of holding bank accounts in the United States. Adding to these other issues, migrants are skeptical about holding an account at a U.S. bank with enough assurance that their assets will not be lost in case they are repatriated.

Table 3.2. IADB Survey of Remittance Recipients

Survey question	% responding "very interested"
Are you interested in a savings account in a bank?	55
Are you interested in a health or life insurance for you and your family?	48
Are you interested in a credit to finance a small business?	40
Are you interested in a mortgage to buy or build a house?	46
Are you interested in a credit to finance university education for yourself or a family member?	34

Source: IADB.

In light of these issues, Honduran financial institutions are in a favorable position to offer services to remittance senders and recipients and address the opportunity to lessen the financial inclusion gap.

Strategies for Financial Inclusion

Financial institutions in Honduras have recognized the opportunity for financial inclusion of migrants in the United States and remittance beneficiaries in Honduras in the last few years. Rapid increases in the number of Honduran migrants over the last decade have led to unprecedented surge in remittances. Most remittances arrived as cash-to-cash transactions. This type of business—cash in, cash out—did not interest most banks. Fortunately, this situation changed as remittances grew larger, recipients wealthier, expectations about quality of services higher, and competition among remittance service providers fiercer. Over time financial institutions perceived the positive impacts of financial inclusion of remittances recipients (Figure 3.2).

The predominant use of formal channels for remittances creates an amicable environment for financial inclusion, albeit at low levels. Most Honduran migrants do not open bank accounts prior to leaving their country and are unlikely to open one in the United States given their undocumented status. Nevertheless, unlike other remittance corridors, in the U.S.-Honduras corridor 92 percent of remittances are channeled through financial institutions.[2] Conditions are favorable in the Honduran banking system to support financial inclusion of remittances receivers and migrants.

Financial institutions in Honduras have adopted different strategies to invite remittance senders and receivers to be clients. These methods depend on the general attitude of a financial institution toward the market, the level of available information, the use of technology, and regulatory aspects in Honduras and in the United States. The financial institutions leverage their unique market positioning and roles in the remittance market.

Figure 3.2. Evolution from Remittances to Financial Inclusion

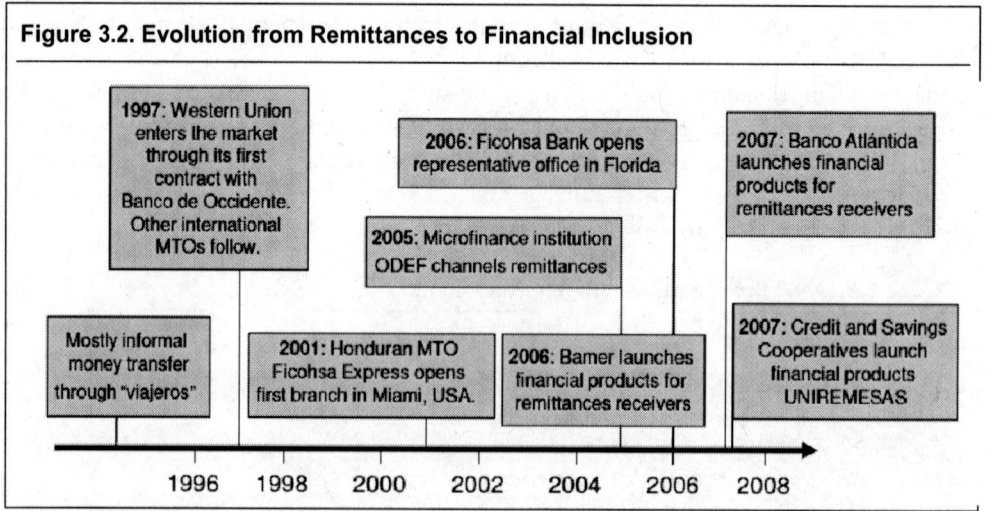

Source: Elaborated by authors based on data publicly available data and interviews with financial institutions.

Financial Inclusion of Recipients

Direct deposit of remittances to beneficiaries' bank accounts is a preliminary stage to cross-sell remittances with other products; yet direct deposits account for only about 12 percent of disbursed remittances across the entire market.[3] Opening a checking or savings account is a defining moment in a bank-client relationship; at this point the customer becomes better known to the institution and can be offered a range of services suited to their individual needs. Direct deposit is a relatively basic feature that allows remittances to enter bank accounts of the beneficiary as soon as they pass the banking or MTO networks. Unlike the cash-to-cash transaction, direct deposits require no physical appearance at the bank branch.

Direct remittance deposits allow the beneficiary to earn interest automatically on the money and create incentives for saving a portion of remittances that is not used for current needs. The financial institutions, on the other hand, can use the deposits for lending operations, and so the bank-multiplier effect is automatically triggered. An important precondition for direct deposit of remittances is an automatic relay between the network (banking or MTO) and the disbursing agent (bank branch, bank agent, or MTO). This facility is not available between Western Union and Banco Occidente, the largest remittance-paying partnership in Honduras.

Minimum deposit requirements create barriers to bancarize the poor. Certain banks in Honduras require a minimum balance of 1,000 lempira to open and maintain an account. Such a balance might be as high as some remittance transfer and may discourage a recipient from opening a bank account. Banks with more efficient cost management and amicable policies now require a minimum balance of 500 lempira, an amount more appealing to prospective customers who are just entering the banking system.

Most financial institutions in Honduras focus on the receiver as the gateway to financial inclusion. Banking services interested in attracting remittances receivers tailor products based on remittances and expanded services. A checking or savings account for receivers is packaged with benefits tailored to the beneficiary's need such as multi-currency accounts, repatriation insurance, private healthcare consultation, international ATM, debit and credit cards, and cell-phone notification of incoming remittance transfers and deposits (Box 3.1). Financial institutions are aware that location is also a factor. Thus banks with national coverage have lately expanded to less populated towns through a network of bank correspondents, which provide limited services but are a cost-effective alternative to a fully fledged bank branch.

Banks and other financial institutions are positioned to increase financial education as a precondition to greater financial inclusion. Beneficiaries in Honduras may have never benefited from financial services prior to the first remittance transfer from a relative in the United States. Remittance withdrawals become a regular interaction and opportunity to tap into other financial services. Educating the new consumer is beneficial to the financial institution as well. The FDIC has a well-established financial education program initiated in 2001, *Money Smart*. This comprehensive financial education curriculum was designed to help individuals outside the financial mainstream develop financial skills and positive banking relationships. The FDIC has far exceeded its original commitment to reach one million consumers. The FDIC continues to work diligently to form alliances with other major entities, including financial institutions; national NGOs; community- and consumer-based groups; and federal, state, and local agencies to promote financial education.[4] Box 3.2 illustrates another approach taking place in Honduras to adapt and upgrade financial literacy.

Several financial institutions have pioneered the use of online banking and mobile phones to expand access to services by receivers. Since 2007, clients of *Banco Ficohsa* and the mobile operator TIGO have been able to access financial services using mobile phones. The services include credit card payments, quotes of loans, balance inquiries of checking and savings accounts, in-coming and out-going transfers, notifications of credit card payments, and loan repayment.[5] Another cell phone service, BAC Movil provided by BAC Bamer, offers similar features but with any mobile phone operator.[6] Remittance beneficiaries of Banco Atlantida (both account and non-account holders) also receive automatic notifications of deposits or incoming remittance transfers.[7] Box 3.3 describes an innovate mobile phone banking service in the Philippines.

Box 3.1. From Remittances to Financial Inclusion—Initiatives by Banco Atlantida

Recent initiatives demonstrate Banco Atlantida's interest in increasing the financial inclusion of remittance beneficiaries. The commercial bank, with over 150 branches and 100 agent networks, offers an account, Cuenta de Ahorros Remesas Atlantida, dedicated to remittance receivers. The account offers several value-added services such as direct deposit of remittances in lempiras or U.S. dollar, preferential interest rates, a debit card, some free ATM withdrawals, medical consultation per phone or private hospitals, repatriation insurance, and insurance against ATM fraud.

Moreover, Banco Atlantida expanded its national coverage by a network of 100 banking correspondents, similarly to financial institutions in other Latin American countries. Other services include withdrawals (with and without a debit card), balance inquiries, and payments. For remittance receivers with previously opened accounts, agents offer a convenient way to use a range of services at a number of often remote locations.

Source: Staff interviews, http://www.bancatlan.hn/.

Box 3.2. BAC BAMER's Life Cycle Model for Financial Inclusion

BAC BAMER, a commercial bank with 100 branches and 180 agents, has seven years experience working on remittance-based financial inclusion with a strong focus on the beneficiaries. Through its 18 partnerships with MTO companies, the bank supports a strategy to select new MTOs based on their interest in financial inclusion, for instance, on the technical possibility and willingness to deposit remittances directly to the migrants or beneficiaries account in BAC BAMER Honduras. The bank runs surveys among remittance beneficiaries and senders on a bi-annual basis and designs remittance strategies on annual basis.

In 2006 BAC BAMER launched a remittance-based savings account, which provides value added services like an assistance plan on health and repatriation, an ATM card for Honduras, as well as easy access to credit cards and credit products.

As a new strategy in 2008, BAC BAMER developed a life-cycle model for financial inclusion based on the following four steps:

1. Remittances kiosk and remittances based savings account (ATM card). In order to provide more remittances receivers with access to their savings account BAC BAMER is piloting an initiative for financial education of remittances receivers through specialized remittances kiosk in selected branches throughout Honduras. These provide specialized information on financial products and services, as well as assistance on questions regarding remittances.

2. Housing loan. After holding remittance-based savings account for a certain time, beneficiaries will be assessed as to their eligibility of a housing loan.

3. Additional loan. At this stage, beneficiaries can access additional loans, for example, to start a small enterprise or strengthen a business.

4. Web- and card-based remittances with a prepaid cash to card facility. This service will be internationally available and also applies to people before they leave to the United States and those Hondurans protected by a temporary protected status.

Source. Elaborated by authors based on interviews with BAC BAMER representatives.

Box 3.3. New Ideas on Mobile Banking and Remittances in the Philippines

Globe Telecom is a leading mobile network operator in the Philippines and is working with CGAP to create ecosystems or mini-economies with multiple locations for people to transact with GCASH, their mobile banking service, via SMS messaging.

Through intensive marketing, targeted customer education, and rapid sign-up and accreditation of retailers, the project will bring mobile phone-based payments and money transfer services for the first time to 3 predominantly low-income rural provinces. The three pilot provinces are expected to reach 80,000 GCASH users.

For more information, search "mobile phone" at www.cgap.org.

With regard to financial inclusion, credit and savings cooperatives offer advantages that have particular benefits for development at local level. Credit and savings cooperatives in Honduras play an important cyclical role by turning remittances into deposits, leading to deposits into lending activities. The credit and savings cooperatives paid out nearly US$70 million in remittances in 2007. Unlike commercial banks, cooperatives in Honduras are connected through their federation and usually do not have a national coverage but rather are tied to specific, often rural, communities. Thus portions of remittances channeled and saved with a cooperative—even in the poorest communities of the country—can generate productive investments within those same communities. In other words, credit and savings cooperatives can allow the bank multiplier effect to work and contribute to economic development at the local level.

Credit and savings cooperatives in Honduras develop lending products tailored to specific development needs of local communities. Agricultural and small and medium enterprise loans are not ordinarily supported by commercial banks. But credit and savings cooperatives have developed a special line of products for remittances receivers called UNIREMESAS and offer many individual services to migrants based on their knowledge of local communities. One credit and savings cooperative, for instance, maintained ongoing business relationships with its members—migrants abroad—to advise on remittance as well as investment and lending products for future businesses.

Financial Inclusion of Senders

Focus on the remittance sender as an entry point to financial inclusion is an alternative strategy of Honduran financial institutions. It is not uncommon that senders make decisions on how to use remittances on their own or the beneficiary's behalf. Many Hondurans left their country, however, without opening a bank account and may not have the required documents to open accounts at U.S. banks. Fewer Hondurans with bank accounts, relative to other migrant groups in the United States, might explain the lack of dedicated services of U.S. institutions for Honduran migrants. As well, the lack of Honduran accounts might be considered weak justification for the cost of opening a dedicated federal- or state-licensed bank, or even branches by a Honduran institution in the United States.[8] Table 3.3 provides a summary of financial strategies for the bancarization of senders and receivers of remittances.

Other financial products are offered to Honduran bank account holders in the United States. Financial institutions in Honduras establish channels of communication with the migrants during their stay abroad. Migrants stay in touch with bank or cooperative branch managers by phone, call center, online, or during return visits to Honduras. Box 3.4 gives an example of one's bank's approach to working with migrants in providing housing and car loans, international ATM, and debit and credit cards in Honduras.

Migration through social networks leads to clusters of migrants from one specific place of origin to another specific place of destination. These clusters make it easier to map migrants and establish contact with them. The transnational bridge approach presented in Chapter 4 offers considerable potential for financial inclusion especially for financial institutions operating on the local level.

Table 3.3. Summary of Honduran Financial Strategies to Bancarize Senders and Beneficiaries

Type	Honduras	Interim stage	U.S.	Advantages	Challenges
Basic model (without financial inclusion)	*Financial institution:* Pays out remittances in cash *Focus:* Receiver	MTO	MTO agent collects money from migrants.	Commissions and favorable exchange rate. Possible earnings from float. Additional service makes financial institution attractive.	Agencies fill up with people. There is a need for lots of cash.
Promotional model	*Financial institution:* Creates strategy for cross-selling of traditional products and services or develops specialized products and services. *Focus:* Receiver	MTO	*MTO agent:* MTOs are selected to facilitate financial inclusion. *Financial institution:* Invests in advertisement to become paying agent	Remittances facilitate contact with potential new clients. Broaden client base. Make use of direct deposit.	Ads in the U.S. (and Honduras) are costly. Information needed on migrants preferences concerning financial products. Level of financial literacy relatively low.
Representative model	*Financial institution:* Looks for a closer relationship with the sender rather than the receiver of remittances. *Focus:* Sender	MTO	*Financial institution:* Establishes strategic alliance with partner organization in U.S. (another financial institution, NGO, MTO, etc.), which functions as front desk and carries out financial education, processing of account applications, promoting products and services among migrant community, etc.	Trust level and access to migrant community rises through permanent representative in U.S. Products and services are being sold to senders, who have proven to often be the ones to make the decisions about.	Finding an adequate partner can be difficult. Very good information about migrant clusters needed, so as to assure that partner is present in those regions.
Representation model	*Financial institution:* Looks for a long-term positioning in migrant market in the U.S. *Focus:* Sender	MTO or bank-to-bank transfer	*Financial institution:* Has its own representation in the U.S.	Additional services can be offered (shipping telecommunications, travel agency, etc.).	If banking license required, this is extremely costly. For global players with presence in U.S. and Honduras, it requires learning how to gain trust and access to migrant's community.
Combination model	*Retail store:* Pays out remittances offering direct connection to the purchase of goods. *Focus:* Receiver	MTO	*Financial institution:* Functions as strategic partner for retail store and provides financial products, i.e. costumer credit cards and debit accounts to remittance receivers.	Cross-selling of consumer goods with consumer credit based on remittances.	Need of large infrastructure

---

Box 3.4. Banco Ficohsa's Approach to Bancarization of Migrants in the United States

In April 2006, Banco Ficohsa obtained a license from the Florida Office of Financial Regulation to open the bank's representative office in Coral Gables, Florida. Unlike a typical bank branch, the representative office only facilitates communication between clients in the United States and Banco Ficohsa in Honduras. The representative office liaises with corporate clients interested in investments in Honduras, as well as retail clients, documented and undocumented Honduran migrants and other nationals.

While it cannot open U.S. accounts for migrants (requiring federal or state bank license), the representative office provides a way to bancarize migrants with Banco Ficohsa in Honduras. For instance, it offers an opportunity to apply for a current account, a debit card, a mortgage loan for the purchase or construction of a house in Honduras, and a car loan in Honduras. The migrant's pattern and history of remittance transfers to Honduras is considered in taking the loan decision. Banco Ficohsa processes the loan (customer due diligence, credit analysis, and so forth) in Honduras.

Migrants can open and manage their accounts in Honduras using a 1-800 number in the United States. At the same time, clients can use the service of the Ficohsa Express MTO to remit cash for savings, debit card payments, or mortgage installments directly to their accounts at Banco Ficohsa in Honduras. It is estimated that about 65 percent of Ficohsa Express clients hold a bank account with Banco Ficohsa in Honduras.

Source: Staff interviews, http://www.ficohsa.com/2007/banco/rep-office.html.

Key Findings and Policy Recommendations

Both U.S. and Honduran authorities should promote financial inclusion and expansion of access to financial services by migrants and their families.

Promote inclusion and expand access with proper identification. Currently, U.S. authorities do not take positions on use of consular identification cards by undocumented migrants.[9] Many commercial banks in the United States accept consular identification cards as a form of identification for migrants. In order for Honduran migrants to enjoy this privilege, the Honduran authorities should develop capacity to issue secured consular identification cards for Honduran migrants in the United States.

Box 3.5. FDIC Money Smart—A Financial Education Program

The Federal Deposit Insurance Corporation (FDIC) initiated a national financial education campaign in 2001 by launching Money Smart, a comprehensive financial education curriculum designed to help individuals outside the financial mainstream develop financial skills and positive banking relationships. The FDIC has far exceeded its original commitment to reach one million consumers. The FDIC is continuing to work diligently to form alliances with other major entities, including financial institutions, national non-profit organizations, community- and consumer-based groups, and federal, state, and local agencies to promote financial education.

Source: Federal Deposit Insurance Corporation. http://www.fdic.gov/consumers/consumer/moneysmart/index.html.

Raise awareness of need for financial education. Honduran consulates, financial institutions, and migrant communities should work with ongoing efforts by regional FDIC offices to raise awareness and conduct basic financial education among Honduran migrant communities.

Improve capacity of the public sector. In order to implement the above policy recommendations, the Honduran authorities should improve the capacity of their consulates in the United States and capacity for issuing secured national identification cards and consular identification cards. This will facilitate undocumented migrant workers to have access to financial services, if these cards are considered secured by financial institutions. The Honduran authorities should enhance the capacity of Honduran consulates to serve the large Honduran migrant population in the United States in other areas of need.

Notes

[1] *Remesas en Centroamérica* (IADB, FELABAN, Bendixen & Associates 2007).

[2] In the Italy-Albania remittance corridor, for instance, 60–70 percent of remittances are transferred using informal means, primarily cash couriers (Hernández-Coss et al. 2006).

[3] CEMLA and MIF (2007). Another source suggests that around 19 percent of all remittances are deposited into bank accounts (Orozco 2008).

[4] Search "Money Smart" on www.fdic.gov.

[5] "FIOCEL, la nueva banca por cellular" (*La Tribuna*, April 29, 2009), www.latribuna.hn/news.

[6] BAC BAMER website, www.bac.net/honduras/esp/banco/personal/perserele.html.

[7] See Chatain et al. (2008) for discussion on recent developments in mobile phone financial services.

[8] Such strategy is pursued by financial institutions in host countries of other diasporas, for instance, by PKO BP in Poland and mBank (BRE Bank S.A.) in the United Kingdom following the recent rapid increase of Polish migrants in that country.

[9] Mexican and Guatemalan consulates in the United States issue their consular identification cards to their own nationals.

CHAPTER 4

Development Impact of Remittances in Rural Honduras: Transnational Economy, Networks, and Diaspora Engagement

This chapter discusses the potential of rising transnational business development relations between rural Honduras and their correspondent migration destinations in the United States. It presents three principal subnational corridors between the United States and Honduras, compares and analyzes different patterns and dynamics between the subnational corridors, and provides a short description of three case studies—Garífuna, Intibucá, and Olancho—of ongoing transnational initiatives and growing interconnectedness of people on the move.[1]

The analysis presented in this chapter shows how poor and rural regions underserved by current national policies get involved in transnational business, cultural exchange, social involvement, and political dialogue. An understanding of subnational remittance corridors and their widespread transnational networks is necessary to efficiently target remittance-based outreach and financial inclusion strategies.

Rising Transnational Economy in Rural Honduras

International migration in rural Honduras gives rise to complex social and economic interaction between the regions of origin and destination of migrants and creates development opportunities that go beyond the transfer of remittances. Honduran migrants play an important role as partners in the social, economic, and political development of their home communities. Migrants' financial contributions to community development, returning migrants' investments in local private sector, export of nostalgic products, courier services, and informal market of migration and remittances have been instrumental to the rising transnational economy in rural Honduras. Migrants already account for an estimated US$1.2 billion of business by purchasing home country goods, travelling to Honduras, getting a mortgage, running a business, calling home, and adding value to migrant organizations.[2]

Collective Remittances and Community Development

Migrants organize themselves to promote the development of their community at home through highly scattered collective remittance initiatives. Migrants from Intibucá (Western Honduras) invest in local infrastructure projects in their hometowns. Garífuna migrants (from the north coast)[3] promote community development on both sides, at home and among diaspora, and voice strategies to advocate for their interests. Their contributions toward community projects resemble collective remittances, a common practice to channel funds for community projects among many migrant diasporas worldwide. Yet, this communal strategy is not broadly known in Honduras; and thus far takes the form of disconnected initiatives in many regions on varying scales.

In southern Intibucá in Western Honduras, 72 identified community projects have been financed by migrants. The aggregated investment by the migrants over the last 7 years in electricity, drinking water, road access, health, education, community centers, soccer fields, and churches is estimated at about US$650,000. Migrants' investments attracted an additional US$150,000 in matching funds from local municipalities, NGOs, and donor agencies; 40 communities (about 20,000 people) benefited from these investments (Figure 4.1). The average amount of a project was US$10,000. In the case of electricity projects, the contribution varies between US$1,000–2,000 per household. Electricity projects are the most common and according to the public electricity company, 55 percent of their clients in Intibucá have financed their access to electricity through remittances Box 4.1 describes how the collective remittances worked to bring electricity to Intibucá.[4]

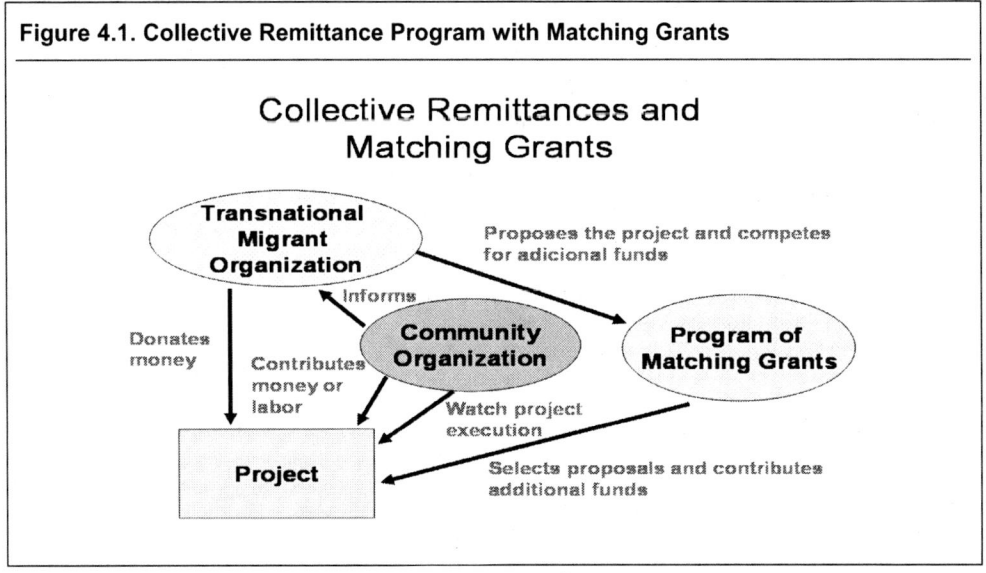

Figure 4.1. Collective Remittance Program with Matching Grants

---

Box 4.1. Collective Remittances at Work in Intibucá

Community members organized a committee to design an electricity project for Intibucá. The project idea was given to a viajero who brought it to a migrant leader in the United States, who then organized a corresponding committee for the collection of funds. The requested money was carried back by the viajero or sent home by each contributing migrant. The electricity committee at home contracted a certified engineer to elaborate a feasibility study on hooking up the remote community to the public electricity system. The public electricity company analyzed the project. After receiving approval the electricity committee bought the necessary equipment and hired the specialized workers. The community helped out doing the less-specialized work. At completion, the public electricity company checked and approved the installation and had the new clients sign the transfer of the installations to the property of the state. To cover any unforeseen problems in the newly public-owned installation, the new clients deposit into a security fund. Intibucá households now have electric lights. The new clients must pay their monthly electricity bills by travelling four to six hours to the closest bank branch.

Sources: PROMYPE/GTZ and IOM (2007).

The Business of Returning Migrants

Returned migrants from Intibucá and Olancho invest their savings and new skills in local businesses. In the town of La Esperanza, the departmental capital, commercial and financial center of Intibucá, 11 percent of businesses are financed by remittances of returned migrants. They run hotels, restaurants, retail stores, as well as transportation and communication enterprises, and are engaged in import-export trading, courier services, real estate, construction, and tourism. Most of the returned migrants (67 percent) have worked on a regular migration status in the United States and continue to travel back and forth for business. Fifty-seven percent complement their remittance savings with a loan to start a business; 73 percent are planning to enlarge their business and 37 percent already manage two or more enterprises. On average businesses employ three to four people, of which 80 percent are not family members. In Olancho, Eastern Honduras, returned migrants make up 3 percent of the population. And 6 percent of businesses in rural communities depend on their investment. Most of these businesses are managed by women. Returned migrants show a rising degree of entrepreneurship in transnational mobility; after some years of migration experience, many of them initiate so-called back-and-forth careers, which manifest in different business activities.

Many of the returning migrants return at some point to the United States for temporary work and leave a family member in charge of their business. The Catacamas Chamber of Commerce observes a pattern of migration, return migration, business set-up and staff training, which is then often followed by another migration cycle. Strong social networks and flexible job opportunities permit such a pattern, which is also strengthened by close ties with U.S. employers who communicate with returned migrants when they are needed. Some of these back-and-forth travelers start to work as viajeros.

The Business of the Courier Services—Viajeros

Viajeros deliver specialized transnational door-to-door export-import courier services between certain regions in Honduras and their correspondent migration networks in the United States. These transnational business people have between 100 to 1,000 clients and deliver in both ways (mainly by air cargo) documents, letters, photos, videos, medicine, traditional food (nostalgic products), remittances, toys, clothes, household and entertaining utilities, furniture, and so forth. Viajeros charge US$5 per pound for the transport of goods and 4–5 percent for delivery of cash. Their price structure is the same everywhere in Honduras. Each journey to the United States leaves them a net profit of around US$1,000. The BCH estimates that channeling of cash using this informal way amounts to 6 percent of total remittance flows (refer to Chapter 2).

The majority of viajeros form a one-person, informal business. Many viajeros are experienced migrants who are well connected to transnational networks. Their basic obstacles are noncompliance with quality and safety standards of the exported nostalgic products, visa regulation, lack of financial history, and principally lack of operation as a registered business. Only few decide to register. However, some viajeros are capitalizing upon their entrepreneurship and becoming professionally managed and legally registered international courier businesses.

Viajeros are for many migrants an important source of information about products, services, prices, and investment opportunities back home. The comparative advantage of the viajero is credibility within the migrant community because, most often, they come from the same community. Their lucrative job (at least by local standards) is seen by many of their clients more as a favor than a service. Viajeros also exchange project documents of collective remittances between the hometown community and migrant leaders or Honduran Hometown Associations (HTA).

Viajeros are an important yet unofficial link between migrant communities and their hometowns in the U.S.-Honduran remittance corridor. "In El Salvador, thousands of individuals earn a living in this manner and have even formed an association (National Association of Couriers and Culture) with approximately 5,000 members."[5] More understanding of the magnitude this phenomenon is needed in Honduras in order to make viajeros participants of transnational development initiatives.

Export of Nostalgic Products

Migrants in the United States create market demand for nostalgic products from Honduras such as traditional food items and local handicrafts (Box 4.2). Compared to

---

Box 4.2. Export of Casabe-bread to Garífuna Community in the United States

Fifteen Garífuna communities in the north coast exclusively organized by women with local Wagabari Distribution and partners in the United States launched a business alliance for trade of nostalgic goods. The exported product is seasoned casabe-bread made out of the traditional staple, the yucca plant.

The distribution in the U.S. market is organized through New Horizon Investment Club, a Garífuna entrepreneurship in New York, whose objective is to combine economic performance with social benefits for Garífuna communities in Honduras. This case is an illustrative example of the business potential on a specific transnational bridge approach between 300 producer families in rural Honduras and the migrant market in New York. Its advantage lies in the efficiency of targeting the consumer in the United States where approximately 100,000 Garífuna live and demand casabe.

Source: Interview with Wabagari Distribution.
For more information see: http://www.newhorizoninvestclub.com/home.

El Salvador, the market for the export of nostalgic Honduran products is still insignificant (accounting for 2 percent of total exports to the United States) but has an extraordinary potential for growth. Export of nostalgic products from El Salvador grew by 60 percent after the Central America–Dominican Republic–United States Free Trade Agreement became effective in 2004. [6] Nostalgic products include cheese, corn products, yucca and banana, sweets, processed beans, coffee, and soups. The staples are cheese (44 percent) and corn products (33 percent). The 2004 estimates were 20 percent (or US$23 million) of the national production of all nostalgic products were exported, principally to El Salvador and the United States.

The traditional handcrafting way of doing business is the principle limitation for effective competition in a rising migrant market in the United States. Despite the limited production output, 88 percent have registered their business in Honduras, out of which 62 percent do not have any sanitary registration for their products and 46 percent do not have adequate packaging.[7] Most are smaller enterprises with less than five employees; their distribution channels utilize migrants' transnational networks and their courier services. Fifty-one percent receive some kind of technical assistance and 89 percent started their business without any financial services.[8]

Migration Dynamics in Three Sub-Transnational Bridges between the United States and Honduras

Honduran migrants in the United States tend to cluster in areas with a high presence of peers from their home communities. *People from Intibucá* tend to group in the Greater Washington, DC Area, *People from Olanchos* (primarily from Sta. María del Real) in Miami (around West Palm Beach), and the Honduran *Garífuna* people traditionally settle in the Bronx or Brooklyn, New York. Complementary to their transnational networks, migrants establish specific subnational remittance corridors. A subnational remittance corridor and their related transnational networks of migrants and families create a transnational bridge—people, goods, money, and information moving/travelling back and forth between the place of a migrant's origin and the destination.

Box 4.3. The Case of Olancho—Florida Transnational Bridge

The Olancho Case is based on a regional household survey of four municipalities accounting for 232,057 inhabitants. Out of these communities are 22,824 migrants, 72 percent from rural and 28 percent from urban areas.

The principal municipality is Catacamas with 103,000 inhabitants. Of the 15 financial institutions that offer remittances services, 68 percent are paid out by only one financial institution. Remittances are channeled to 35 percent of households and correspond to 34 percent of the region's GDP. About one-third of the Olanchano migrants moved to Miami, another 17 percent went to New York, and 11 percent to Boston.

At the municipal level, migration to one specific U.S. site is much more pronounced than aggregated data shows. For example, 44 percent of the migrants of the municipality of Campamento moved to certain places in Miami demonstrating the importance of social transnational networks.

Sources: Authors based on research carried out by RDS and UNA.

Underlying the bilateral remittance corridor between Honduras and the United States, different subnational remittance and migration corridors developed over time with distinct migration dynamics and economic impacts. The understanding of the differences and their related development potentials are important for policy initiatives and the private sector, especially for financial institutions. The three sub-transnational bridges—Olancho–Florida; Intibucá–Greater Washington, DC area; and Northern Honduras (Garífuna)–New York—were selected for this study, not just because of their common features, but because they present important differences that are explained by social, ethnic, geographic, and historic factors. They present a vivid panorama of the evolution of migration dynamics over time and their different levels of transnationalism, even in a small country like Honduras.

The Olancho Case

The Olancho case is representative for the most recent migration patterns today in Honduras. Migration was primarily triggered by the aftermath of the disastrous Hurricane Mitch; the majority of migrants are young men from rural areas with irregular migration status (Box 4.4).

Box 4.4. The Case of Intibucá—Greater Washington, DC Area Transnational Bridge

Approximately 200,000 people live in rural Intibucá, which has the second lowest Human Development Index of communities in Honduras. Six financial institutions offer remittance services, but one financial institution dominates with 55 percent of the market share. Remarkable for Intibucá is the rural outreach of cooperatives in municipalities with highest migration rates. About US$30 million in remittances were sent to Intibucá in 2006. This is 10 times more than the total annual budget of all 17 municipalities that incorporate this department. About 20 percent of the households receive remittances on a regular monthly basis, although these are distributed unevenly throughout Intibucá. It is estimated that more than the half of Intibucá migrants in the United States live in the Greater Washington, DC, area.

Source: Authors based on research carried out by PROMYPE/GTZ.

Compared to other cases, *coyotes* play a particularly active role in the Olancho–Florida transnational bridge. Seventy-four percent of the migrants are paying for human smuggling schemes, making that a very lucrative business (refer to Chapter 1).[9] These facts demonstrate that the more recent the migration tradition of a region—under the given U.S. admission policies—the more migrants rely on informal migration and remittance services.

Due to their recent migration history, the Olanchanos in the United States are only sporadically engaged in community support of their hometowns. Only very few cases could be identified where migrants' collective remittances support community projects at home. And it is expected that with time, these still feeble transnational bridges between the Diaspora and hometown communities will strengthen.

The Intibucá Case

The Intibucá case is representative for the typical Central American migration pattern of the past, triggered by civil wars and counter insurgency. International migration in Intibucá started in the south of this department about 20 years ago, in those municipalities that border with El Salvador. This region hosted refugee camps and suffered many of the consequences of the Salvadoran civil war between 1980 and 1992. The first Intibucanos left their communities under the chaotic circumstances of the civil war in the neighboring country and entered the United States as Salvadoran refugees. Later the emigration trend spread to other parts of the department (Box 4.5).

Box 4.5. The Case of the Northern Coast (Garífuna)—New York Transnational Bridge

For the past 200 years, the Garífuna population has maintained its traditional culture in the coastal areas of Belize, Guatemala, Honduras, and Nicaragua, subsisting through the cultivation of yucca and other crops, and through fishing. There are 48 Garífuna communities in north Honduras, but no reliable data on the Garífuna population is available. Scholars estimate the population around 300,000; and Garífuna-advocacy NGOs, including La Organización de Desarrollo Étnico Commuitario (ODECO), calculate that Afro-Caribbean descendants comprise 10 percent of the population in Central America. The largest contingent of Garífuna is thought to reside in 21 communities in the department of Colón. Santa Rosa de Aguan estimates 80 percent of households with migrants.[a] Based on a representative household survey in 2004, 27 percent of the Garífuna households in Honduras receive regular remittances; 53 percent receive more than US$500 and 47 percent receive less than US$500.[b] The single largest source of these remittances is from New York.

Source: Authors based on interviews carried out with ODECO (La Ceiba, May 2008).
a. PROMYPE/GTZ and DED (2004:15).
b. ODECO (2002).

The number of community projects financed by migrants through collective remittance schemes in Intibucá is impressive. The south of Intibucá, once considered one of the most marginalized and poorest regions of Honduras—like all neighboring regions along the Salvadoran frontier—is a vivid example on the positive development impact of international migration where migrants' investment in the infrastructure of their hometowns is overruling most public investment and makes a tangible difference to people's living standards.

Although the Intibucanos followed similar migration footpaths, contrary to Salvadorans, they did not institutionalize their Honduran Hometown Association and prefer much more informal ways to organize community support for their hometowns. The high degree of informality of Honduran migration in the United States is a major obstacle to scaling up development incentives. Amending existing collective remittance schemes with matching funds, for instance, would have a significantly larger impact on local development initiatives.

The Garífuna Case

The Garífuna people have been migrating to North America since at least the 1930s.[10] Today, migrant remittances represent a key resource, which preserves the Garífuna culture in Central America (Box 4.6).[11] First migration waves were triggered by employment with Merchant Marines and United Fruit Company, initially to New Orleans. It continued with family reunifications and tripled due to Hurricane Mitch in 1998. Now, there are about 100,000 Garífuna in the United States (one-third of its native population), of which at least 60,000 are in the Bronx, making it the largest Garífuna community in the world. Other migration destinations of Garífuna are Miami, Los Angeles, and London. Most Garífuna in the United States reside legally. This was possible due to well-established family networks built by more than 60 years of migration.[12] The majority of Garífunas left for the United States for education and employment and continue to travel back and forth; but the majority return home at some time in their life.

The Garífuna maintain the largest and most effective diaspora network of all Honduran migrants in the United States. Fifty Garífuna organizations have been identified; some promote culture and ethnic identity, others get involved in the development of their hometowns.[13] The Garífuna have a proven history of supporting hometown development through collective remittances. Half of the 48 communities have a corresponding Hometown Association in the United States, albeit the majority acting on an informal basis. Since the 1970s, Garífuna migrants have been concerned with development of their villages through infrastructure development, mutual aid societies, and the celebration of Garífuna culture. They collect monthly contributions of their members and raise funds through different activities (benefit events, raffles, and visits to casinos).[14]

Summary of the Three Cases

Lessons from all three cases suggest that understanding subnational remittance corridors and their underlying transnational migrant networks helps design and implement more efficient outreach and financial inclusion yet from a low initial scale. A subnational perspective helps turn informal migration patterns to local development opportunities, builds trust, and engages key stakeholders at a local level (Table 4.1). The challenge is to scale up development initiatives by leveraging unique features of transnational networks and a supportive overall national policy framework.

Box 4.6. Criteria for the Emergence of Hometown Associations

Manuel Orozco (2007) points out four criteria that determine the emergence of Hometown Associations and their involvement in the country of origin: (a) the level of community consciousness of migrants, especially of their elite; (b) the level of outreach of the homeland government; (c) the perception of migrants by the society in the homeland; and (d) the relationship between the governments of the country of origin and destination. The following table shows criteria for the emergence of transnational engagement of the Honduran Diaspora according to their peculiar migration patterns:

	Olancho	Intibucá	Garífunas	Honduras
Level of community consciousness of Honduran migrants	Low	Medium	High	Low
Honduran (or local) government encouraging diaspora identification	Not existent	First tentative	Punctual tentative via Garífuna organization in Honduras	Low and with political bias toward party affiliation
Perception of migrants and its impact by society in Honduras	Negative	Negative	Negative	Negative
Relationship between Honduran and U.S. government	-	-	-	Focused on the annual prolongation of TPS Permits

Source: Authors.

Table 4.1. Migration Patterns and Stakeholder for Subnational Outreach Initiatives

	Migration pattern	Stakeholder
Olancho	Irregular migration	Courier
Intibucá	Collective remittances	Community leaders
Garífunas (North Coast)	Social networks	HTA

Source: Authors.

Table 4.2 summarizes the three cases in Garífuna, Intibucá, and Olancho.

Table 4.2. U.S.-Honduras Transnational Bridges: Summary of Three Cases

	Olancho	Intibucá	Garífuna (North coast)
Trigger/push-pull reasons for migration	Hurricane Mitch	Civil war and counter insurgency in neighboring country	Job opportunities by Merchant Marine and international commerce
Initial arrival date	72% after 2000	During 1980	Since the 1930s
U.S. concentration	30% in Florida	50% live in Greater Washington, DC	60% in New York
Receiving households	35%	20%	27%
Social network and voice	1 HTA	Many activities on an informal basis	22 informal HTA, 4 registered HTA, and several advocacy NGOs with international activities and outreach
Collective remittances	Only few projects identified	72 identified	Long tradition of mutual aid and emergency help
Remittance services	68% channeled by one financial institution	55% by one financial institution	No data
migration status	Recent migrants are 95% on an irregular status	No data available	60% with regular migration status
Return migrants, investment and entrepreneurship in rural communities	Remittances investment in private sector represents 6%	Remittances investment in private sector represents 11%	No data
Courier services/ viajero	Very important	Important	Important
Exportation of nostalgic products	Informal	Informal	Formal
Outstanding characteristics	Courier services	Collective remittances	HTA

Source: Authors.

Capacity building among both migrants and home communities are necessary preconditions for increasing the developmental impact of transnational bridges. The Olancho case shows that migration is overwhelmingly organized through informal channels, making the local courier system essential for outreach policies. In Intibucá, networks developed by migrants and community leaders at home with neighboring El Salvador led to impressive investments in local infrastructure. The Garífuna case, however, presents the most advanced form of transnationalism due greatly to their established migration history and ethnic bonds. For outreach policies, it would be best to address first the Garífuna organization in the United States and Honduras. The case of the Garífuna shows higher concentration of migrants in destination and regular migration status; more elaborated social networks, HTA, and advocacy organizations; longer tradition of mutual and emergency aid; collective remittances projects; formalization of the export of nostalgic products; and more impact of remittances and investment at home.

Compared to neighboring countries the circumstances for the emergence of Honduran Hometown Associations are not favorable. In contrast to organizations established by other Diaspora communities in the United States, not much is known about the HTA activities (Box 4.7). And just 6.5 percent of Honduran migrants in the United States belong to a Hometown Association.

Honduras faces an opportune time to strengthen transnational bridges with the assistance of many stakeholders. Only recently have financial institutions started to look for ways to reach out to both senders and beneficiaries of remittances in order to cross-sell financial products and promote financial inclusion. Local political authorities are looking for ways to strengthen the involvement of migrant leaders in community development. Local, national, and international NGOs offer matching funds for migrants' financial contribution to local development. These are examples of a growing awareness and interest of different actors to build on strategies that involve both sides, the region of origin and destination of migrants, and that could be visualized for development purposes as a transnational bridge.

Key Findings and Policy Recommendations

The above analysis and presentation of case studies has shown how local economies of semi-rural areas in Honduras are connected to the places of destination of their migrants through transnational bridges. Economic impacts go beyond remittances and include the movement of people and the exchange of goods, money, and information. Transnational bridges open up opportunities to promote local economic development in high migration areas through measures that involve migrants, their families, and transnational entrepreneurs.

Create matching fund programs for migrant's community investments. Other countries in the LAC Region have created public or private matching fund programs that complement migrants' investment in their home community's social infrastructure. Migrant associations usually register with their consulates and compete for extra funding through their project proposals. Beyond the positive effect of additional social infrastructure in migrants' home communities, these programs help to connect migrant associations to initiatives of local development and can ultimately turn them into dialogue partners.

Create migrant friendly investment policies at the local level. Some migrants plan to go back to their hometowns and invest their savings to create an income for themselves and their family. Others might be interested in helping a family member with their business idea. Local development agencies, municipalities, or others could help these migrants develop investment ideas and business plans by providing information on topics such as the following: the local economy (prices, competition, lack of products or services, investment opportunities, and so forth) business courses, legal and fiscal requirements, and sources of additional financing. Additionally, fiscal incentives could be an adequate measure to attract migrant's investments back home.

Strengthen export of nostalgic products. Migrants in the United States create demand for locally produced goods, especially foodstuffs and other typical items, which often cannot be bought abroad or, when available, do not taste the same. Local goods create a nostalgic bond with the hometown. The demand for locally produced goods presents a new and growing market for local producers who often already send their products

---

Box 4.7. The Transnational Bridge: Toward a Development Methodology

The concept of a transnational bridge—bringing together senders and beneficiaries of the same origin—was the marketing strategy of one financial institution to promote their products and services through social corporate investment in education. This financial institution targeted the remittance market on the migration network between the towns of Talanga (Francisco Morazan) and Gerona (Spain) based on its corporate social responsibility approach.

Institutions interested in identifying and working in a transnational bridge can apply the following steps, which present a first rough approach towards a methodology of working on transnational bridges. Further research and most of all practical experiences are still necessary to identify best practices when working on economic development and with migrants and their families on the local level.

Step 1—Understanding the big picture
The first step is the understanding of the economic, social, and political map of local key actors in rural areas, and their personal and professional relationship to migration, remittances, and local development. After gathering basic information, the objectives of the project and the opportunities of participation are presented to local authorities. Finally a working compromise is agreed on and tasks are assigned to participants of the project.

Step 2—Investigating by involving local actors
With the help and direct involvement of mayors, private sector, dignitaries, NGOs, and universities, information on the economic implications of international migration and remittances in a particular region is gathered through participatory diagnostics, small surveys, and semi-structured interviews with local resource persons.

Step 3—Communicating the results
Massive communication of the results to the participants and the public is decisive for trust building and acceptance of the approach, specifically for two reasons: (a) to make the approach more accountable to the public and (b) to put international migration and its development opportunities on the agenda of local actors. Social marketing is necessary to break walls, open minds, and change attitudes.

Step 4—Getting started (the first initiatives)
After investigating and communicating the issues of migration, remittances, and local development, the first initiatives get started in those areas, which seem most promising as a quick result—like facilitating the communication between senders of remittances and financial institutions. Initiatives only work if participants see them as a business opportunity.

Step 5—Reach out to the migrants
Steps one to four are understood as trust-building measures to connect two sides of a transnational bridge with the purpose of showing the development opportunities of remittances and migration. The last step directly involves migrant leaders in the country of destination in order to incentivize capacity building and the institutionalization of Honduran Hometown Associations as legitimate and visible counterparts of community development and partners for private-public projects.

Source: Authors.

to the United States through viajeros. Formalizing and amplifying these exports are challenges. For example, local producers might need help in getting sanitary registration, export licenses, information on necessary permits and transport, and how to commercialize their products in places where migrants live.

Contact and incorporate talent abroad. Connecting highly skilled migrants to development initiatives on the local level creates opportunities for knowledge transfer and innovation. Identifying talents and creating networks of these intrinsically motivated people is a strategy applied by some countries to connect their business and scientific communities to top-level knowledge and provide them with contacts, as well as mentoring or internship programs.

Notes

[1] Research on the transnational economy has been pioneered by Manuel Orozco and the 5Ts (family remittance transfers, tourism, transportation, telecommunication, and nostalgic trade) of the transnational economy model. For more information see Orozco (2005).

[2] See Appendix Table A.25, Transnational Activities of Honduran Migrants.

[3] Garífuna are people of African and American Indian descent that live mainly along the Caribbean coast of northern Central America (PROMYPE/GTZ and DED 2004).

[4] PROMYPE/GTZ and IOM (2007). PROMYPE-GTZ prepared a documentary about how projects with collective remittances function. The documentary advocates public and private institutions to contribute to a program of matching grants.

[5] Andrade-Eckhoff and Silva-Avalos (2003: 27).

[6] Interview with the Vice-Ministry of Foreign Affairs of El Salvador (December 2007).

[7] In 2004 the Foundation for Investment and Export (FIDE) identified 117 businesses dedicated to the production of nostalgic products, but only 23 of them exported to the United States or Central America. Sixty percent did not have a registered product name; almost half did not use a specific product packaging; and 44 percent of the products had milk derivates, which did not comply with international sanitary standards: all obstacles for export.

[8] FIDE (2004).

[9] RDS and IDRC (2007b: 57).

[10] PROMYPE/GTZ and DED (2004).

[11] Gonzalez (1979).

[12] PROMYPE/GTZ and DED (2004: 16).

[13] Ávila (2006a, 2006b).

[14] The first HTA, *Asociación Unión Corozaleña ASUNCOR* (1969), engaged in fundraising for water, electricity, and education projects. New York accounts for 22 informal HTA and 4 registered HTA like *Organización de Damas Limoneras, Travesia Nueva Ola, Jóvenes de Funda,* and *Pro-Desarrollo de Aguan.* Others are *Bajamar, Las Aquellas,* and *Unión Corozaleña.* Other organizations like the Garífuna Coalition USA offer social services to the Garífuna immigrant community in the Bronx. *Hondureños contra el Sida* is engaged in capacity building on both sides of the transnational bridge to improve awareness of sexually transmitted diseases. The New Horizon Investment Club offers technical assistance to hometown communities like Tornabe and Miami that acquired 7 percent of shareholder value of the tourism project, Los Micos Beach and Resort. Besides the solidarity of diaspora organizations in the United States, advocacy NGOs in Honduras are engaged in promoting the interest of Garífuna on an international level.

CHAPTER 5

Key Conclusions and Proposed Roles of Stakeholders

This chapter summarizes key conclusions of the report. In addition, it summarizes the recommendations that followed each chapter and suggests the roles of national and subnational policy makers for improving regulatory framework and implementations, for increasing the efficiency of remittance in the U.S.-Honduran remittance corridor, for increasing the developmental impact of remittances through greater financial inclusion, and for strengthening transnational bridges between diaspora communities and their home towns.

Conclusions

There are both positive and negative implications of the profiles of Honduran migrants. Because they arrived in the United States at younger ages than other Latin American migrants, some of them have had opportunities to finish school in the United States, pushing average education levels higher than other migrants. On the other hand, the negative side includes lower wages, lack of identification card due to their age of departure from Honduras, and the lack of identification preventing them from using banking services.

Remittances are significant for the Honduran economy. They represented 25.6 percent of GDP in 2006. Although the nominal recorded inflow of remittances was US$2.6 billion, which makes Honduras the 9th largest recipient in the LAC Region. The significance of the remittances is massive because of the size of Honduras' economy. The United States plays a significant role given that over 90 percent of remittances originate there. Recent economic stagnation, in particular the housing and construction sectors, appears to have lowered the remittance growth rate, signaling a challenge to sustainable inflows of remittances to Honduras.

The significance of remittances to Honduras has attracted private sector engagement in the U.S.-Honduras remittance corridor and the level of competition has been rising. In the United States, international MTOs, Honduran MTOs, and some regional MTOs have large market shares in the corridor. In Honduras, on the other hand, banks have large shares where remittances through MTOs may channel through banks as agents. Recently, microfinance institutions began to enter the market, seeking niche opportunities such as local positioning and good reputations among clients. Even with rising competition, distribution networks in rural areas are limited despite credit and saving cooperatives trying to capture demand. Informal transfer operators, viajeros, exist but only have about 6 percent of total remittances according to the BCH. Their advantages include quickly and easily bridging two specific locations between the United States and Honduras. Because of rising competition, their significance to the market has been declining.

There is lack of clarity in some existing regulations and lack of regulation in the remittance market in Honduras. This has created different rules for various remittance businesses. Banks and other regulated financial institutions are supervised by CNBS and required to comply with all laws and regulations as financial institutions. On the other hand, pure remittance companies are not regarded as financial institutions and exist as a regular business entity. Lack of clarity in regulations has allowed Honduran banks to offer accounts for Honduran migrants in the United States, although interviews with the Honduran authorities suggest that physical presence is required in opening an account.

A few financial institutions have initiated efforts to expand financial services in order to include remittance recipients who are not account holders. A few banks opened a representative office and established a U.S.-based MSB, partnering with international and regional money service businesses. Also, banks have introduced banking products linked with remittances. However, minimum balance requirements for bank accounts appear to have reduced their attractions to unbanked remittance recipients.

Productive use of remittances, including development of financial services, can be an important relief for macroeconomic management. Remittances have relatively been a stable source of foreign exchange inflows with important economic benefits. However, further increases could create real appreciation pressures and lead to crowding out the export sector. This can create challenges because Honduras' exports sectors (maquila, tourism, and agriculture) play an important role for growth and employment. Thus, fostering financial services that could help channel more remittances into productive uses (such as by improving financial services associated with remittances' flows) will be an important challenge to help promote growth in the future.

Transnational bridges such as hometown associations, transnational businesses, and economies have seen successes. Migrants in the United States, communities at home, and coordinating groups have achieved targeted community development with remittances. Migrant groups in the United States have also opened doors for transnational businesses such as exporting nostalgic goods from Honduras to the United States. Neither of those activities has involved action by the national government. While national policies and strategies are important for macroeconomic policy, migration, regulatory reform for remittance markets, transnational bridges through remittances, and businesses have seen specific achievements of development. However, only a few migrant groups have developed a sense of community in the United States. The majority of Honduran migrants do not form communities in the United States and in particular the embassy and consulates have not played a role in developing communities. As a result, these successes are only observed in a few communities.

Recommended Actions for Key Stakeholders

In addition to the policy recommendations laid out at the end of each chapter, the role of stakeholders is key to policy development and implementation. Improvements in the remittance market, financial inclusion, and development of a robust transnational economy require the collaboration of a number of stakeholders in both countries and on all levels. Table 4.5 summarizes the recommendations and suggests which stakeholders are critical to their implementation.

Table 4.5 Summary of Recommendations: Proposed Stakeholder Actions

P Policy   **A** Action   **C** Collaboration

Recommendation	Honduras				United States			
	Federal gov.	Regional & local gov.	FI	NGO & civic society	Federal gov.	State gov	RSP	Diaspora
Migration & remittance								
Engage in dialogue and exchange experience with neighboring states on migration and remittance issues	A	C	C	C				
Re-evaluate the TPS to a temporary worker program	P & A				P & A			
Improve and collaborate on data collection of migration and remittances	A		C	C				
Develop payment systems infrastructure	P & A		C					
Regulate money transfer companies.	P & A		C					
Clarify regulatory requirements and compliance.	A		C					
Develop a monitoring / supervisory framework.	P & A		C					
Better harmonize and coordinate state regulations and examinations of MSBs in the U.S.					P	P	C	
Strengthen the capacity of Honduran consulates in the U.S.	A		C		C	C	C	C
Financial inclusion								
Promote financial literacy	P & A	P & A	A	A	P & A	P & A	A	A
Public sector capacity building								
Capacity development of Honduran Consulates	A							
Consulates support Honduran community	A			A				
Capacity development of identification issuance	P &A				C		C	
Transnational economy								
Create matching fund programs for collective remittances	P & A	P & A	A				A	A
Create migrant friendly investment policies at the local level		P	A					A
Strengthen export of nostalgic products	P	P	A	C	P	P		A
Contact and incorporate talent abroad	A	A		C				C

Appendix

Table A.1. Price Evolution of the "Coyote Business" Over Time

Time	US$
Before 1997	3,800
1997–01	4,200
2002–05	4,400
2006–07	6,000
2008	Over 6,000

Source: RDS 2005 and 2007; UNA 2005; and interviews by PROMYPE/GTZ in Intibucá, Olancho, and Atlantida in 2007 and 2008.

Table A.2. Growth Rate of Central American Migrants to the United States, 1990–2000

	U.S. Census 2000			Mumford		
	1990	2000	%	1990	2000	%
Total Hispanics	21,900,089	35,305,818	61	21,900,089	35,305,818	61
Total Central Americans	1,167,584	1,491,493	28	1,266,314	2,517,465	99
Honduras	131,066	217,569	**66**	142,481	362,171	**154**
El Salvador	565,081	655,165	16	583,397	1,117,959	92
Guatemala	268,779	372,487	39	279,360	627,329	125

Source: http://www.census.gov and http://www.s4.brown.edu/cen2000/HispanicPop.

Table A.3. Central American Migrants according to American Community Survey in 2006

Salvadoran	1,371,666
Guatemalan	874,799
Honduran	490,317

Source: http://www.census.gov/acs.

Table A.4. Migrants per Period and According to Household Quintiles

Income Quintile	Percentage		
	Until 1989	1999–2003	2004–06
1	6.1	8.2	8.9
2	14.3	14.6	17.2
3	20.9	19.1	20.6
4	25.5	24.2	24.5
5	32.8	33.3	28.1
Ignored	0.6	0.6	0.6

Source: BCH (2007b: 49). Based on INE-household survey in 2006.

Table A.5. Percentage Household Quintiles Receiving Remittances, 2004–06

	Quintile 1	Quintile 2	Quintile 3	Quintile 4	Quintile 5
Total	6.7	14.8	20.2	23.2	30.7
Tegucigalpa	11.9	11.3	16.2	18.2	23.9
San Pedro Sula	15.2	14.3	15.9	18.6	25.6
Middle sized urban Center	13.4	16.2	20.1	23.8	34.4
Rural Areas	5.1	14.8	21.9	26.5	36.7

Source: BCH (2007b: 49). Based on INE-household survey in 2006.

Table A.6. Income of Quintiles of Remittances for Receiving and Non-Receiving Households (in lempiras)

	Quintile 1	Quintile 2	Quintile 3	Quintile 4	Quintile 5
Total	870	2,386	4,329	7,513	25,76
Remittance receiving	1,052	2,420	4,353	7,598	23,965
Non-remittance receiving	857	2,380	4,323	7,487	26,563
% income difference	23%	2%	1%	1%	−10%

Source: BCH (2007b: 53). Based on INE-household survey in 2006.

Table A.7. Destinations according to *IADB/FELABAN: Comparing CA, 2007*

	Percentage			
	United States	Canada	Europe	Others
Honduran	76	1	16	7
Salvadoran	78	9	7	6
Guatemalan	90	7	2	1

Source: IADB/FELABAN: Comparing CA, 2007.

Table A.8. Destination of Central Americans in the United States According to 2000 U.S. Census

State	Hondurans	% of total Hondurans in U.S.	Salvadoran	Guatemalan
Florida	41,229	19.0	20,701	28,650
New York	35,135	16.2	72,713	29,074
California	30,372	14.0	272,999	143,500
Texas	24,179	11.2	79,204	18,539
New Jersey	15,431	7.1	25,230	16,992
Louisiana	8,792	4.1	1,127	2,093
North Carolina	8,321	3.8	8,679	5,966
Virginia	7,819	3.6	43,653	10,000
Illinois	5,992	2.8	7,085	19,790
Georgia	5,158	2.4	8,497	10,718
Massachusetts	5,125	2.4	15,900	11,437
Maryland	4,067	1.9	34,433	8,304

Table A.9. Socioeconomic Statistics of 2007 American Community Survey

	Total population	Mexican	Honduran	Salvadoran	Guatemalan
Migrant Population	301,621,159	29,166,981	430,504	1,474,342	872,334
Male (%)	49.3	52.7	53.1	53.0	57.7
Female (%)	50.7	47.3	46.9	47.0	42.3
Median age (year)	36.7	25.8	33.9	29.2	27.8
Under 18 years (%)	24.6	36.5	8.3	29.6	28.5
Family households (%)	66.8	80.9	76.8	83.7	79.1
Female householder (no husband present) (%)	12.5	16.9	**23.5**	19.6	17.5
Now Married (%)	50.2	49.4	43.7	45.0	44.9
School Enrollment	79,329,527	9,185,631	52,028	389,915	217,310
Elementary School (%)	40.5	49.5	32.7	46.4	47.3
Education attainment for Population 25 years and over: Less than high school diploma (%)	15.5	45.8	**50.4**	53.5	54.3
High school graduate (%)	30.1	**27.9**	**27.5**	24.9	23.4
Some college or associate's degree (%)	26.9	17.7	12.5	13.8	13.7
Bachelor's degree (%)	17.4	6.2	**7.3**	6.0	6.6
Graduate or professional degree (%)	10.1	2.4	2.3	1.7	2.0

(Table continues on next page)

Table A.9 (continued)

	Total population	Mexican	Honduran	Salvadoran	Guatemalan
Foreign born	38,059,555	11,629,457	430,504	968,133	607,231
Naturalized U.S. citizen (female) (%)	16,181,883 (53.3)	2,515,696 (48.5)	94,229 (**59.0**)	277,603 (53.9)	135,283 (51.7)
Entered before 1990 (%)	42.9	37.0	24.7	43.5	31.8
Entered 1990 to 1999 (%)	29.4	32.6	**34.3**	30.2	28.7
Entered 2000 or later (%)	27.7	30.4	**40.9**	26.3	39.6
Speak English less than "very well" (%)	8.7	40.9	**72.6**	56.8	60.7
Commuting to work by car (drove alone) (%)	76.1	66.8	52.3	61.9	52.6
Employment Status: in labor force (%)	64.8	68.4	74.7	76.3	76.5
Occupation: Services (female) (%)	16.7 (37.9)	24.4 (30.9)	30.1 (**48.7**)	32.2 (47.2)	32.6 (48.3)
Occupation: Construction (male) (%)	9.7 (17.5)	18.3 (28.8)	**30.4 (47.9)**	19.0 (30.7)	22.3 (31.5)
Private wage and salary workers (%)	78.6	84.8	86.8	86.5	87.0
Self-employed (%)	6.7	5.8	9.2	7.8	8.7
Median household income (US$)	50,740	39,742	**36.521**	43,633	41,352
Households with earnings in last 12 month (%)	80.3	91.0	94.0	96.4	95.8
With Social Security income (%)	26.9	13,5	6.3	6.6	6.1
per capita income (US$)	26,688	13,823	18,151	15,189	14,380
Poverty Rates all families (female householder) (%)	9.5 (28.2)	20.0 (39.7)	**22.9 (39.6)**	14.2 (29.3)	18.9 (36.2)
Renter-occupied housing units (%)	32.8	48.8	**66.5**	52.3	65.9
Owner-occupied housing units: monthly owner costs as % of household income: 30 percent or more	30.5	43.6	62.4	65.2	62.0
Owner characteristics: Median Value (US$)	194,300	175,300	228,800	325,600	287,800
Owner characteristics: monthly owner cost with a mortgage (US$)	1,464	1,444	1,750	1,890	1,835
Gross rent as percentage of household income in the past 12 months (30 percent or more)	45.6	51.0	55.6	52.4	52.9
Vehicles available (%)	91.3	91.0	80.8	88.7	82.8

Source: ACS 2007.

Table A.10. Immigrant and Non-Immigrant Status of Hondurans in United States

- Hondurans obtaining legal permanent resident status between 1997 and 2006: 63,128 (average 6,223 a year)
- Hondurans naturalized between 1997 and 2006: 39,607 (average 3,960 year)
- Non-immigrant admissions to United States (I-94 only): for Hondurans around 100,000 a year
- Non-immigrant temporary worker admissions (I-94 only) in 2006: Total 2,579 (664 workers in specialty occupations, seasonal agricultural workers 5, seasonal non-agriculture worker 841, entertainers 302, intra-company transfers 225)

Source: Yearbook 2007, Office of Immigration Statistics.

Table A.11. Frequency of Sending Remittances

	Percentage	
	Once a month or more often	Over a month
Honduras	54	38
Salvador	61	36
Guatemala	59	40

Source: MIF/FOMIN, IADB, FELABAN, Miami 6.11.2007

Table A.12. Migrants Sending Remittances Home

	Percentage		
	Less than 1 year	Between 1–3 years	More than 3 years
Honduras	24	35	36
Salvador	6	48	43
Guatemala	13	30	55

Source: MIF/FOMIN, IADB, FELABAN, Miami 6.11.2007

Table A.13. Average Amounts of Remittances Transfers

	US$
Honduras	225
El Salvador	300
Guatemala	240
Central America	240

Source: MIF/FOMIN, IADB, FELABAN, Miami 6.11.2007

Table A.14. Remittances Sent Home by Migrants in the United States

Average US$	Honduras	El Salvador	Guatemala
50 (50 or less)	17	18	12
100 (51–100)	34	35	36
150 (101–150)	10	15	11
200 (151–200)	21	17	20
250 (201–250)	0	2	4
300 (251–300)	5	5	6
350 (301–350)	1	1	1
400 (351–400)	1	1	1
450 (401–450)	0	1	1
500 (451–500)	3	2	3
More than 500	3	2	2
No response	3	2	2

Source: MIF/FOMIN. Receptores de Remesas en América Latina: El Caso Colombiano, (2004:.12).

Table A.15. Comparison of RSP Market Shares in U.S.-Central America Corridors (2004)

Transfer Institution	Honduras	El Salvador	Guatemala
Post Office	5	5	6
Western Union	43	26	33
MoneyGram	16	3	7
King Express	0	0	35
Gigante Express (Courier)	3	15	2
Banks	18	34	6
Personal Courier	9	13	3
Other	4	3	4
No response	2	0	4

Source: MIF/FOMIN. Receptores de Remesas en América Latina: El Caso Colombiano, 2004.

Table A.16. Distribution of Remittances in Honduras: Alliances of Banks and Cooperatives with their Agents, 2007

	Agent	Representatives	#		Agent	Representatives	#
1.-	Banco Bamer	Ria Envia	1	13.-	Banhcafe	Western Union	
	Cuenta con 54 Sucursales a Nivel Nacional	Delgado Travel	2		Cuenta con 38 Sucursales a Nivel Nacional	Giros Latinos	
		Intermex	3			Order Express	54
		Afex	4			FMI El Salvador	55
		BTS	5			Servi Giros	
		Corfinge	6			Dolex	
		Money Link	7				
		MoneyGram	8				
		Giro Express	9	14.-	Elektra	Vigo	
		Giros Latinos	10		Cuenta con 21 Sucursales a Nivel Nacional	Dinero Express	56
		Ria (España)	11			Order Express	
2.-	HSBC	Giros Latinos	12	15.-	Cooperativa Sagrada Familia	Western Union	
	Cuenta con 57 Sucursales a Nivel Nacional	MoneyGram	13		Cuenta con 25 Sucursales a Nivel Nacional		
		Vigo	14				
				16.-	Cooperativas UNIRED		
3.-	Ficohsa	Ficohsa Express	15		Cuenta con 47 Sucursales a Nivel Nacional		
	Cuenta con 52 Sucursales a Nivel Nacional	Dolex	16		Guadalupe	Vigo, MoneyGram, Uniteller	
		Girosol	17		Elga	Vigo, MoneyGram, Uniteller	
		Uniteller	18		Apaguiz Ltda	Vigo, MoneyGram, Uniteller	
		Telecom	19		Ceibeña Ltda	Vigo, MoneyGram, Uniteller	
		MoneyGram			Coompol Ltda	Vigo, MoneyGram, Uniteller	
4.-	Cuscatlan	Corfinge			Intibucana Ltda	Vigo, MoneyGram, Uniteller	
	Cuenta con 14 Sucursales a Nivel Nacional				Juticalpa Ltda	Vigo, MoneyGram, Uniteller	
					Ocotepeque Ltda	Vigo, MoneyGram, Uniteller	
5.-	Atlantida	MoneyGram			San Antonio	Vigo, MoneyGram, Uniteller	
	Cuenta con 111 Sucursales a Nivel Nacional	Ban Comercio	20		Maria Claret Ltda	Vigo, MoneyGram, Uniteller	

Agent	Representatives	#	Agent	Representatives	#
	Occidente Corp	21	Taulabe Ltda	Vigo, MoneyGram, Uniteller	
	Elexa	22	Usula Ltda	Vigo, MoneyGram, Uniteller	
	Ria Envía		San Marqueña Ltda	Vigo, MoneyGram, Uniteller	
	Order Express		La Fronteriza Intibucana	Vigo, MoneyGram, Uniteller	
	Vigo		Fraternidad Pespirense Ltda	Vigo, MoneyGram, Uniteller	
			Rio Grande Ltda	Vigo, MoneyGram, Uniteller	
6.- Occidente	Western Union	23	Arsenault Ltda	Vigo, MoneyGram, Uniteller	
Cuenta con 114 Sucursales a Nivel Nacional	Cambio Universal / Casa de Cambio Lempira / Dennis Castillo	24	Fe y Esperanza Ltda	Vigo, MoneyGram, Uniteller	
	Giros Latinos / Vigo		La Nueva Vida	Vigo, MoneyGram, Uniteller	
	Honduras Express	25	Chorotega	Vigo, MoneyGram, Uniteller	
	Servicio de Envíos	26	La Caceenp	Vigo, MoneyGram, Uniteller	
	Intermex		La Talanga	Vigo, MoneyGram, Uniteller	
			17.- Servi Giros	Vigo	
7.- Banpais	Western Union		Manejan 22 remesadoras	Delgado Travel	
Cuenta con 58 Sucursales a Nivel Nacional			1-877-628-3610	Mateo Express	57
8.- Ficensa	Western Union		Cuenta con 03 Sucursales a Nivel Nacional	Pronto Envíos	
Cuenta con 23 Sucursales a Nivel Nacional				Uno	58
9.- Credomatic	Inter America	27		Ría	
Cuenta con 32 Sucursales a Nivel Nacional	Via America	28		Rapido Envíos	59
	Vermounth	29		Remesas Agil	60
	Vanuys	30		Ibero America	61
	Vigo			Colombia Express	62
				Girosol	
10.- Continental	Banco de El Comercio	31	España	Unigiros	63
Cuenta con Sucursales a Nivel Nacional				CheckPoint	64

Agent	Representatives	#	Agent	Representatives	#
			Londres	LCC Trans Sending	
11.- Banco de Los Trabajadores	Servicio de Envíos		Canada	Multi Express	65
Cuenta con 17 Sucursales a Nivel Nacional			16	Otros sin identificar claramente	66
12.- Banco Uno	Aval Envíos	32			
Cuenta con 16 Sucursales a Nivel Nacional	Quisqueyana	33	Laitano Service		67
	El Camino Transferencia	34	18.- Cuenta con 03 Sucursales a Nivel Nacional	Tegucigalpa	
	Merchnts Bank of California	35		San Pedro Sula	
	International Money Transmisión	36		La Ceiba	
	La Nacional	37			
	Banco Hipotecario Dominicano	38	Gutierrez Cargo y Familia		68
	Giromex	39	19.- Cuenta con 60 Sucursales a Nivel Nacional	Tegucigalpa 60 oficinas	
	Intertransfers	40			
	Efectivo Money Transfer	41			
	Enramex	42			
	Telecomm	43			
	Multienvios	44			
	RD Money Trasfer	45			
	Telegiros	46			
	Capital Money Transfer	47			
	Transcard	48			
	Moneytel	49			
	Aval Card Costa Rica	50			
	Banco Uno Costa Rica	51			
	First Remit	52			
	LCC Trans Sendin	53		Source: Based on interviews with market participants	

Table A.17. Remittance Fees Paid for Sending US$200 from the United States to Honduras (2008)

Organization	Transfer type	Method	Last Updated	Fee	Speed	Account Needed
Sole Provider International	Cash transferred electronically	online	Last Month	$3.00	Instant	Yes
Bank of the West	Cash transferred electronically	agent	Over a Month	$4.00	3–7 days	No
Xoom.com	Credit card to e-wallet	telephone	Over a Month	$4.99	4 days	No
Cathay Bank	Cash transferred electronically	agent	Over a Month	$6.00	3–7 days	No
Pacific Western Bank	Cash transferred electronically	agent	Over a Month	$6.00	3–7 days	No
MoneyGram	Bank account to check	agent	Over a Month	$9.99	Instant	No
Bank of America	Cash transferred electronically	agent	Over a Month	$10.00	3–7 days	No
eMoneyGram	Bank account to check	telephone	Over a Month	$10.00	3–5 days	No
Xoom.com	Bank account to check	telephone	Over a Month	$10.99	4 days	No
iKobo	Credit card to credit card	telephone	Over a Month	$11.00	Instant	Yes
Western Union	Bank account to check	agent	Over a Month	$11.99	Instant	No
Western Union	Bank account to check	telephone	Over a Month	$11.99	Instant	No
Western Union	Bank account to check	branch	Over a Month	$11.99	Instant	No
Wells Fargo	Cash transferred electronically	agent	Over a Month	$12.00	3–7 days	No
eMoneyGram	Bank account to check	telephone	Over a Month	$14.00	Instant	No
Bank of America	Cash transferred electronically	agent	Over a Month	$15.00	3–7 days	No
Cathay Bank	Cash transferred to Bank account	agent	Over a Month	$25.00	3–5 days	Yes
Citibank	Cash transferred to bank account	telephone	Over a Month	$30.00	1–24 hours	Yes
HSBC	Cash transferred to bank account	agent	Over a Month	$30.00	3–5 days	Yes
Wells Fargo	Cash transferred to bank account	agent	Over a Month	$32.00	3–5 days	Yes
Bank of America	Cash transferred to bank account	agent	Over a Month	$35.00	3–5 days	Yes
Bank of the West	Cash transferred to bank account	agent	Over a Month	$35.00	3–5 days	Yes
Citibank	Bank account to check	branch	Over a Month	$40.00	1–24 hours	No
Pacific Western Bank	Cash transferred to bank account	agent	Over a Month	$40.00	3–5 days	Yes

Source: www.sendmoneyhome.org visited on July 11, 2008.

Table A.18. Remittance Fees Paid in the U.S.-Honduras Corridor (2008)

Amount		MTO 1 Fee	%	MTO 2 Fee	%	MTO 3 Fee	%	FI 1 Fee	%	FI 2 Fee	%	FI 3 Fee	%	FI 4 Fee	%
$0.01	$50.00	$13.00	26.0	$12.00	24.0	$4.50	9.0	$40.00	80.0	$65.00	130.0	$50.00	100.0	$35.00	70.00
$50.01	$100.00	$15.00	15.0	$12.00	12.0	$6.00	6.0	$40.00	40.0	$65.00	65.0	$50.00	50.0	$35.00	35.00
$100.01	$200.00	$22.00	11.0	$15.00	7.5	$9.00	4.5	$40.00	20.0	$65.00	32.5	$50.00	25.0	$35.00	17.50
$200.01	$300.00	$29.00	9.7	$20.00	6.7	$12.00	4.0	$40.00	13.3	$65.00	21.7	$50.00	16.7	$35.00	11.67
$300.01	$400.00	$34.00	8.5	$20.00	5.0	$15.00	3.8	$40.00	10.0	$65.00	16.3	$50.00	12.5	$35.00	8.75
$400.01	$500.00	$40.00	8.0	$30.00	6.0	$18.00	3.6	$40.00	8.0	$65.00	13.0	$50.00	10.0	$35.00	7.00
$500.01	$600.00	$45.00	7.5	$30.00	5.0	$21.00	3.5	$40.00	6.7	$65.00	10.8	$50.00	8.3	$35.00	5.83
$600.01	$700.00	$45.00	6.4	$40.00	5.7	$24.00	3.4	$40.00	5.7	$65.00	9.3	$50.00	7.1	$35.00	5.00
$700.01	$800.00	$50.00	6.3	$40.00	5.0	$27.00	3.4	$40.00	5.0	$65.00	8.1	$50.00	6.3	$35.00	4.38
$800.01	$900.00	$50.00	5.6	$50.00	5.6	$30.00	3.3	$40.00	4.4	$65.00	7.2	$50.00	5.6	$35.00	3.89
$900.01	$1,000.00	$50.00	5.0	$50.00	5.0	$33.00	3.3	$40.00	4.0	$65.00	6.5	$50.00	5.0	$35.00	3.50
$1,000.01	$1,100.00	$75.00	6.8	$60.00	5.5	$36.00	3.3	$40.00	3.6	$65.00	5.9	$50.00	4.5	$35.00	3.18
$1,100.01	$1,200.00	$75.00	6.3	$60.00	5.0	$39.00	3.3	$40.00	3.3	$65.00	5.4	$50.00	4.2	$35.00	2.92
$1,200.01	$1,300.00	$75.00	5.8	$80.00	6.2	$42.00	3.2	$40.00	3.1	$65.00	5.0	$50.00	3.8	$35.00	2.69
$1,300.01	$1,400.00	$75.00	5.4	$80.00	5.7	$45.00	3.2	$40.00	2.9	$65.00	4.6	$50.00	3.6	$35.00	2.50
$1,400.01	$1,500.00	$75.00	5.0	$80.00	5.3	$48.00	3.2	$40.00	2.7	$65.00	4.3	$50.00	3.3	$35.00	2.33
$1,500.01	$1,750.00	$80.00	4.6	$80.00	4.6	$55.50	3.2	$40.00	2.3	$65.00	3.7	$50.00	2.9	$35.00	2.00
$1,750.01	$2,000.00	$90.00	4.5	$120.00	6.0	$63.00	3.2	$40.00	2.0	$65.00	3.3	$50.00	2.5	$35.00	1.75

Source: Fees delivered by MTO and FI.

Table A.19. Interest of Households in Financial Products (percent)

	Central America	Honduras	El Salvador	Guatemala
Savings account	53	55	54	50
Life or health insurance	44	48	49	31
Credit for business	38	40	49	21
Mortgage for housing	31	46	34	16
Credit for university education	25	34	25	9

Source: IADB/FELABAN: Comparing CA, 2007.

Table A.20. Reverse Remittances

2005	2006	2007
US$302,000	US$1,148,000	US$10,000 (estimated)

Source: BCH.

Table A.21. Frequency of Immigrant Travel to Home Country

	Percentage		
	Honduran	Salvadoran	Guatemalan
Tree times or more a year	0	1.5	0.9
Twice a year	5.5	5.6	3.7
Once a year	6.8	20.4	4.6
Once every two years	12.3	5.6	3.7
Once every three years	2.7	8.7	0.9
Travel little	12.3	23.5	15.6
Never travelled	60.3	34.7	70.6

Source: Data from Manuel Orozco's 2003–2004 survey of immigrants in New York; Los Angeles; Washington, DC; Chicago; and Miami; administered by Emmanuel Silvestre and Associates.

Table A.22. Deportable Hondurans and Other Aliens from the United States, 2001–06, by Country of Nationality (Office of Immigration Statistics, Yearbooks)

	2001	2002	2003	2004	2005	2006
Honduras	10,803	11,295	16,632	26,555	55,775	33,365
Salvador	11,688	9,209	11,757	19,180	42,884	46,329
Guatemala	7,434	8,344	10,355	14,228	25,908	25,136

Source: Yearbooks, Office of Immigration Statistics.

Table A.23. Deported Hondurans from U.S. Reported by Centro de Atención al Migrante (CAMR) at the Tocontins Airport Tegucigalpa

	2001	2002	2003	2004	2005	2006	2007
Honduras	3,903	6,304	7,105	9,350	18,941	24,643	29,348

Source: Centro de Atención al Migrante (CAMR) at the Tocontins Airport Tegucigalpa.

Table A.24. Deportable Hondurans by Mexican Authorities

	2004	2005	2006
Honduras	73,046	79,006	59,013
Salvador	35,270	42,952	26,930
Guatemala*	93,667	100,630	84,657

Source: Instituto Nacional de Migración, Mexico, Estadísticas Migratorias (2007).
*High rate of apprehensions of Guatemalan is explained through estimated annual 40,000 non-documented temporal workers in southern Mexico.

Table A.25. Transnational Activities of Honduran Migrants

	%	Number	Cost	US$
Call on average 120 minutes (month)	57	342,000	30	10,260,000
Buy home country goods	74	444,000	200	88,800,000
Travel once a year to Honduras	12	72,000	700	50,400,000
(and spend over US$1,000)	43	258,000	1,000	258,000,000
Have a mortgage loan	12	72,000	7,000	504,000,000
Own a small business	4	24,000	7,500	180,000,000
Helps family with mortgage	8	48,000	2,000	96,000,000
Belong to an HTA	7	42,000	200	8,400,000
TOTAL				1,195,860,000

Source: Criteria based on works of Manuel Orozco from the Institute for the Study of International Migration, Washington, DC. Calculation based on estimated 600,000 adult migrants.

Table A.26. HTA and Advocacy-NGO of Garífunas in New York

Period	Number of organizations*	Name of organization and year of foundation
1945–69	siguiente	Carib American Association, 1946 Belice Honduran Assocciation of New York, 1956 Fenix Social Club, 1959 Honduras Football & Social Club, 1965
1970–79	4	Asociación Unión Corozaleña "ASUNCOR", 1969/70 (first HTA)
1980–89	1	Mujeres Garinagu en Marcha (MUGAMA), 1989
1990–99	28	Hondureños Contra el SIDA, 1992 Patronato de Bajamar en New York, HTA, 1996 The Garífuna Coalition USA, 1999, http://Garifunacoalition.org
2000–06	12	Club de Inversión Nuevos Horizontes, (Investment in Turism) 2000, www.newhorizoninvestclub.com Grupo Las Aquellas HTA (Triunfo de la Cruz), 2000

Source: José Francisco Ávila, Organizaciones no lucrativas Garífunas de Nueva York. Working Paper. New York, 2006.
*36 Organizations from Honduran, 8 from Belize, and 6 from Guatemala.

Table A.27. Different Migration Patterns in Three Subnational Remittances Corridors

	Olancho	Intibucá	Garífuna (North coast)
Trigger/push-pull reasons for migration	Hurricane Mitch	Civil war and counter insurgency in neighboring country	Job opportunities by Merchant Marine and International Commerce
Date	Migrations picks up after 2000 (72%)	During 1980	1930
Concentration in the United States	30% in Florida (Miami), 17% in New York (New York City), 11% in Massachusetts (Boston), 9% in Missouri (San Luis)	50% live in Greater Washington (27% in Virginia, 17% in Maryland, 13% in New Jersey)	30% of the Garífuna living in the United States, mainly in New York
Receiving households	10% of population lives in the exterior, representing 28% of the households. 35% receive remittances.	20% of households with migrants in the department, but in southern Intibucá 50%. Households with migrants geographically very uneven distributed.	27% of households
Importance of remittances	Remittances paid in 2006 are US$75 million and correspond to 34% of local GDP.	US$30 million in 2006, 10 times higher than the annual budget of the 17 municipalities in the department.	No data.
Social network and voice	Only 1 HTA identified (West Palm Beach).	No HTA identified, but many activities on an informal basis.	22 informal HTA, 4 registered HTA and several Advocacy-NGOs with international activities and outreach
Collective remittances	Only few projects identified.	Every community with a higher rate of international migrants relies on collective remittances, 72 identified.	Long tradition of mutual aid and emergency help. Many communities receive collective remittances.
Remittance services	75% Western Union, 8% MoneyGram, 8% banks, 3.4% viajeros, 5.4% no response. 68% are channelled through one bank out of 15 financial institutions that pay remittances.	6 financial institution pay remittances, but 55% by 1 financial institution. 80% of the senders condition the money sent to housing, real estate or education. 50% of the receivers have an account in a financial institution.	No data

(Table continues on next page)

Table A.28 (continued)

	Olancho	Intibucá	Garífuna (North coast)
Migration status	9% are residents, 2% citizens, 9% with TPS, 17% based on some type of permission (tourist, working, study visa), 55% are undocumented, and 8% no answer. But recent migrants are 95% on irregular basis and 74% employed and paid a so-called "coyote" who charges between US$5,000 and $7,000.	No data	Estimated 60% with regular migration status.
Return migrants, investment and entrepreneurship in rural communities	3% of the population returned to Olancho. 8 out of 10 emigrants are planning to return. Remittances investment in private sector represents 6 percent.	Visible impact of returned migrant in local economy, 54 business identified only in the small town center of the departmental capital La Esperanza. Remittances are invested in all economic sectors of the department.	Investment of successful migrants in tourism, real estate, local business, import-export, technology, communication, transport.
Courier services/ viajero	Very important	Important	Not important
Exportation of nostalgic products		Only on a informal basis through Viajeros	Two enterprises export Yuca-Chips (Casabe).
Outstanding characteristics	Viajeros.	Collective Remittances.	HTA.

References and Select Bibliography

Agunias, Dovelyn Rannweig, and Kathleen Newland. 2007. "Circular Migration and Development: Trends, Policy Routes, and Ways Forward." Policy Brief. Migration Policy Institute. Washington, DC.

Albanian Government and IOM (International Organization for Migration). 2005. "National Strategy on Migration and National Plan on Migration." Tirana.

Andrade-Eckhoff, Katherine, and Claudia Marina Silva-Avalos. 2003. "The Challenges of Transnational Migration for Local Development in Central America." Working document. FLACSO.

Ávila, José Francisco. 2006a. "Organizaciones no lucrativas Garífunas de Nueva York." Working paper. New York.

———. 2006b. *Memoria de Reunión Comunitaria con Organizaciones Afro descendientes de Centro América en Nueva York.* May 26, 2006.

BCH (Banco Central de Honduras). 2009. Balance of Payments Statistics. Available at: http://www.bch.hn.

———. 2008. *Remesas familiares enviadas por hondureños residentes en el exterior y gastos efectuados en el país durante sus visitas. Informe de encuestas.* Tegucigalpa. January.

———. 2007a. *Consideraciones Sobre Las Remesas Familiares Enviadas A Honduras.* Tegucigalpa. March.

———. 2007b. *El Rostro de las Remesas: Su impacto y sostenibilidad.* Tegucigalpa.

Camarota, Steven B. 2007. *Immigrants in the United States, 2007: A Profile of Americas Foreign-Born Population.* Washington, DC: Center for Immigration Studies Backgrounder.

Cantor, Eric. 2004. "Remittances and Development: Lessons from the Garífuna Transnational Community." GTZ/DED. Honduras.

CEMLA (*Centro de Estudios Monetarios Latinoamericanos*) and Multilateral Investment Fund. 2007. *Remesas Internacionales en Honduras.* Washington, DC.

Chatain, Pierre, Raúl Hernández-Coss, Kamil Borowik, and Andrew Zerzan. 2008. "Integrity in Mobile Phone Financial Services. Measures for Mitigating Risks for Money Laundering and Terrorist Financing." World Bank Working Paper No. 146. Washington, DC.

Cheikhrouhou, Hela, Rodrigo Jarque, Raúl Hernández-Coss. 2006. *The U.S.-Guatemala Remittance Corridor: Understanding Better the Drivers of Remittance Intermediation.* Washington, DC: World Bank.

England, Sarah. 2006. *Afro Central Americans in New York. Garífuna Tales of Transnational Movements and Racialized Space.* Gainesville, Florida: University Press of Florida.

Faijnzylber, Pablo, and J. Humberto López. 2007. *Remittances and Development: Lessons from Latin America*. Washington, DC: World Bank.

Federal Reserve Bank of Chicago. 2006. *Financial Access for Immigrants: Lessons from Diverse Perspectives*. Chicago, Illinois.

FIDE (*Fundación para la Inversión y Desarrollo de Exportaciones*). 2004. *Estudio de Oferta Exportable de Productos Nostálgicos de Honduras*. Honduras.

FONAMIH 2008. *Balance Migratorio: Honduras 2008*. Foro Nacional para las Migraciones en Honduras. Tegucigalpa, Septiembre 2008. Semana del Migrante

Gonzalez, Nancie L. 1979. "Garifuna Settlement in New York: A New Frontier." *International Migration*, Vol. 13, No. 2. Special issue: International Migration in Latin America.

Hernández-Coss, Raúl. 2005. "The U.S.-Mexico Remittances Corridor: Lessons on Shifting from Informal to Formal Transfer Systems." Washington, DC: World Bank.

Hernández-Coss, Raul, et al. 2006. "The Italy-Albania Remittance Corridor: Shifting from the Physical Transfer of Cash to a Formal Money Transfer System." Washington, DC: World Bank and Convergence.

IADB, FELABAN, Bendixen & Associates. 2007. "Remesas en Centroamérica." Presentation. Available at: http:// idbdocs.iadb.org/wsdocs/getdocument.aspx? docnum=1200049.

IMF (International Monetary Fund). 2006. *IMF Country Report No. 06/35*. Washington, DC.

INE (*Instituto Nacional de Estadística*). 2007. *Migración y Remesas Internacionales*. BID. Presidencia de la República. Honduras

———. 2004. *Encuesta Nacional de Hogares sobre Condiciones de Vida (ENCOVI)*. Tegucigalpa. Honduras.

Massey, Douglas S., ed. 2008. *New Faces in New Places: The Changing Geography of American Immigration*. New York: Russell Sage Foundation.

ODECO (*La Organización de Desarrollo Étnico Commuitario*). 2002. *La Comunidad Garífuna y sus Desafios en el Siglo XXI*. Hondurus.

Orozco, Manuel. 2008. *Migración y Remesas en Honduras—Sinopsis de los resultados de la última encuesta sobre remesas y migración realizada en Junio 2008 por Borges & Asociados*.

———. 2007. "Conceptualizing Diasporas: Remarks about the Latino and Caribbean Experience" In: Sorensen, Ninna Nyberg. 2007. *Living Across Worlds: Diaspora, Development and Transnational Engagement*. Geneva: IOM.

———. 2006. "International Flows of Remittances: Cost, Competition and Financial Access in Latin America and the Caribbean—Toward an Industry Scorecard." Inter-American Dialogue. Washington, DC.

———. 2005. "Transnational Engagement, Remittances and their Relationship to Development in Latin America and the Caribbean." Inter-American Dialogue. Washington, DC.

Orozco, Manuel, and Rachel Fedewa. 2006. "Leveraging Efforts on Remittances and Financial Intermediation." IDB INTAL—ITD Working Paper 24. December.

PROMYPE/GTZ and DED. 2004. "Remittances and Development: Lessons from the Garifuna Transnational Community." PROMYPE/GTZ and DED, Honduras.

PROMYPE/GTZ and IOM. 2007. *Puente Transnacional Intibucá—EE.UU. Resultados Preliminares de la primera etapa del proyecto.* PROMYPE/GTZ and IOM, Honduras.

Ratha, Dilip. 2003. "Workers' Remittances: An Important and Stable Source of External Development Finance." *Global Development Finance 2003.* Chapter 7. Washington, DC: World Bank.

Ratha, Dilip, Sanket Mohapatra, and Zhimei Xu. 2008. "Outlook for Remittance Growth for 2008-2010: Growth Expected to Moderate Significantly, But Flows to Remain Resilient." *Migration and Development Brief 8.* Migration and Remittances Team, Development Prospects Group. World Bank. Washington, DC.

RDS and IDRC (Red de Desarrollo Sostenible and International Development Research Center). 2007a. *Proyecto "Impacto de la migración y remesas en la economía local de Olancho."* Área Económica Informe Final. With CRDI—Canada. Tegucigalpa. Honduras.

———. 2007b. *Proyecto "Impacto de la migración y remesas en la economía local de Olancho."* Área Demográfica Informe Final. With CRDI—Canada. Tegucigalpa. Honduras.

———. 2007c. *Proyecto "Impacto de la migración y remesas en la economía local de Olancho."* Área Género Informe Final. With CRDI—Canada. Tegucigalpa. Honduras.

———. 2007d. Impacto de la Emigración y la Remesa en la Economía Local de Olancho *Uso, Ahorro e Inversión de la Remesa / Fuente de Ingreso de los Hogares. Documento de discuisón.*

UNAT-UNFPA. 2006. "Migración, Mercado de Trabajo y Pobreza en Honduras." UNAT-UNFPA, Honduras.

U.S. Census Bureau. 2008. American Community Survey 2007. Available at: http://www.census.gov.

———. 2007. American Community Survey 2006. Available at: http://www.census.gov.

U.S. Department of Homeland Security. 2009. "Estimates of the Unauthorized Immigrant Population Residing in the United States: January 2008." Washington, DC.

World Bank. 2009. "Migration and Development Brief 9." Migration and Remittances Team, Development Prospects Group. March 23. World Bank, Washington, DC.

———. 2009a. Remittance Price Database. World Bank, Washington, DC. Available at: http://remittanceprices.worldbank.org.

———. 2009b. Remittance and Migration Data, Development Prospects Group. World Bank, Washington, DC. Available at: http://go.worldbank.org/QOWEWD6TA0.

———. 2008a. *Migration and Remittances Factbook.* Washington, DC: World Bank.

———. 2008b. Remittance Price Database. World Bank, Washington, DC. Available at: http://remittanceprices.worldbank.org.

World Bank and CPSS (Committee on Payment and Settlement Systems). 2007. "General Principles for International Remittance Services." World Bank and CPSS.

World Bank Prospects Group. 2009. "Revised Outlook for Remittance Flows 2009–2011." World Bank, Washington, DC.

Eco-Audit

Environmental Benefits Statement

The World Bank is committed to preserving Endangered Forests and natural resources. We print World Bank Working Papers and Country Studies on postconsumer recycled paper, processed chlorine free. The World Bank has formally agreed to follow the recommended standards for paper usage set by Green Press Initiative—a nonprofit program supporting publishers in using fiber that is not sourced from Endangered Forests. For more information, visit www.greenpressinitiative.org.

In 2008, the printing of these books on recycled paper saved the following:

Trees*	Solid Waste	Water	Net Greenhouse Gases	Total Energy
355	16,663	129,550	31,256	247 mil.
*40 feet in height and 6–8 inches in diameter	Pounds	Gallons	Pounds CO$_2$ Equivalent	BTUs

green press
INITIATIVE